STRATEGIC RETHINK

STRATEGIC CHOICES FOR A TURBULENT WORLD

In Pursuit of Security and Opportunity

Andrew R. Hoehn | Richard H. Solomon | Sonni Efron
Frank Camm | Anita Chandra | Debra Knopman | Burgess Laird
Robert J. Lempert | Howard J. Shatz | Casimir Yost

RAND CORPORATION

For more information on this publication, visit www.rand.org/t/RR1631

Library of Congress Cataloging-in-Publication Data is available for this publication.

ISBN: 978-0-8330-9692-0

Published by the RAND Corporation, Santa Monica, Calif.

Cover design by Pete Soriano

Preface

This is the sixth and final volume in the RAND Strategic Rethink project. This series has developed conceptual perspectives on how U.S. thinking, institutions, and policies must adapt to the many changes in the international environment. Together, these studies help clarify the strategic choices facing the country in 2017 and beyond.

The first volume, anchored by Ambassador James Dobbins, outlines the foreign policy choices that U.S. policymakers now face in three critical regions—the Middle East, Europe, and Asia—as well as on such problems as counterterrorism, climate change, and cybersecurity.

The second study, on national defense, by David Ochmanek and Andrew Hoehn, demonstrates that the United States suffers a "security deficit" between its stated military strategy and the resources allocated to its defense posture. It outlines what Americans can expect for their defense dollars at four different levels of spending, all of them lower than historic norms. And it argues that the United States must either spend more on its defenses or reduce its global security ambitions.

The third volume, by Hans Binnendijk, assesses the state of U.S. alliances and partnerships, exploring three alternative strategies for managing potential adversaries. It concludes that collaborative engagement, though not without constraints, is the most feasible for the United States. It also recommends a trilateral defense strategy that would feature closer ties among the United States, Europe, and Asia.

The fourth, a Perspective by Ambassador Charles P. Ries, probes the deficiencies in the U.S. national security policymaking and policy implementation systems, offering eight recommendations for reorganizing and improving decisionmaking in an era of rising challenges and shrinking policymaker bandwidth.

The fifth, an assessment of the international economy by Howard J. Shatz, concludes that the United States is likely to maintain the world's largest economy for many years, and that it will benefit from continuing its leadership role in the international institutions it helped to build over the past seven decades. The report argues that the United States should improve these global structures and integrate rising powers, demonstrating to other countries that a U.S.-led economic system remains desirable.

We draw on the research and insights of these five associated studies in this concluding volume, as well as contributions from other RAND colleagues across a range of disciplines.

This study should be of interest to U.S. policymakers and lawmakers, analysts, the media, nongovernmental organizations, and others concerned with the role of the United States and other nations in advancing global security and economic growth.

This project results from the RAND Corporation's Investment in People and Ideas program. Support for this program is provided, in part, by the generosity of RAND's donors and by the fees earned on client-funded research.

Special appreciation is due to the Hauser Foundation for its generous gifts in support of the project, and to Rita Hauser for her encouragement and support of the effort.

The RAND Strategic Rethink Series

Available at www.rand.org/research/projects/strategic-rethink

Choices for America in a Turbulent World
James Dobbins, Richard H. Solomon, Michael S. Chase, Ryan Henry, F. Stephen Larrabee, Robert J. Lempert, Andrew Liepman, Jeffrey Martini, David Ochmanek, and Howard J. Shatz
RR-1114-RC, 2015

America's Security Deficit: Addressing the Imbalance Between Strategy and Resources in a Turbulent World
David Ochmanek, Andrew R. Hoehn, James T. Quinlivan, Seth G. Jones, and Edward L. Warner
RR-1223-RC, 2015

Friends, Foes, and Future Directions: U.S. Partnerships in a Turbulent World
Hans Binnendijk
RR-1210-RC, 2016

Improving Decisionmaking in a Turbulent World
Charles P. Ries
PE-192-RC, 2016

U.S. International Economic Strategy in a Turbulent World
Howard J. Shatz
RR-1521-RC, 2016

Foreword

This is the sixth and final volume of the Strategic Rethink series, during which RAND has pulled together some of its best minds to take a fresh look at America's role in the world—its interests, ambitions, obstacles, and options for a turbulent new era.

The project was born several years ago, at a moment when both the international security situation and the U.S. domestic political mood seemed to be deteriorating sharply. We had war fatigue at home, tumult in the Middle East, increasing tensions with Russia and China, a Salafi-jihadist movement taking root in new lands, and a rising tide of partisanship threatening to paralyze the U.S. ability to conduct a sustained, coherent foreign policy. I asked Ambassadors Richard Solomon and James Dobbins to lead a wide-ranging effort to reexamine America's challenges and its capabilities. What level of international engagement is the American public willing to support, and to what end? How much does America want or need to lead a world where many problems seem to be beyond its ability to control? Is there a "grand strategy" for diplomacy and defense that would match U.S. interests with the resources required to succeed at that strategy? And if not a "grand" strategy, are there other strategic concepts that could align and orient U.S. foreign and domestic policy and generate support from the American public?

The five previous volumes in this series explored key issues that will face the next president in the Middle East, Europe, and Asia; national defense; alliances and partnerships; national security decisionmaking; and international economic policy. This final overview report analyzes how the United States moved from the triumph at the end of the Cold War to the stalemate of today and suggests a range of changes the United States can make to better adapt to this new era of turbulence and uncertainty. The report offers three plausible strategic concepts that the United States might pursue, and evaluates the underlying assumptions, costs, risks, and constraints. It also offers thoughts on how to choose among alternatives.

We do not advocate any one of these strategic concepts. In fact, we assume that policymakers will mix and match as they strive to cope with fast-evolving circumstances and advance changing U.S. interests. As this report points out, changes in course even during a single administration tend to be the rule rather than the exception.

I would offer five considerations for readers as they use this volume to explore, challenge, and develop their own views of America's role in the world.

First, I am deeply concerned that the United States has contracted a disease we might call "truth decay." In politics and beyond, we see the danger that Sen. Daniel Patrick Moynihan once warned of, that people feel entitled not only to their opinions, but to their own set of facts to support those opinions. One goal of the Strategic Rethink series has been to present a rigorous body of facts and analysis on which to have a productive discussion about strategic choices for America.

Second, American voters don't always get what we pay for, but we almost never get what we don't pay for. Each strategic concept in this report comes with an approximate price tag for defense and security spending. Failure to allocate proper resources to a chosen course of action is a recipe for disappointment, debt, or disaster.

Third, much has been written about the limitations on U.S. power, but the authors caution that inaction on the global stage can sometimes be just as costly as action. A careful calibration of ends and means is always a necessity.

Fourth, the analysis highlights the many strengths the United States brings to bear on the global stage, more than any other nation, and none more important than its many friends and allies. No other competitor has the advantage of such a global network of friends and allies; to surrender or squander this advantage would be a catastrophic strategic failure.

Finally, this report analyzes many external threats to the United States. Yet the authors conclude that an internal problem—domestic political dysfunction—is the greatest threat of all. No effective response to any major problem, whether an international trade deal, a major infrastructure project, a new weapon system, or a tax or entitlement reform program, can succeed in the span of a two-year Congress or a four-year presidential administration. Therefore, bipartisan agreements must not only be forged, they must be sustained in order to achieve meaningful, lasting results.

I hope this report will stimulate the long-range thinking and bipartisan policy planning that will be needed to secure and sustain America's place in a turbulent world.

Michael D. Rich
President and Chief Executive Officer,
RAND Corporation

Contents

Figures, Lists, and Tables

Figures

Lists

Tables

Summary

For more than seven decades, the United States assumed leadership in building a liberal international order. It confronted threats from the totalitarian, nuclear-armed Soviet Union and other authoritarian states. After the Cold War, it organized collective responses to challenges to the international order posed by less powerful aggressors, including tyrants who committed grave violations against their neighbors or their own populations. Americans united after the terrorist attacks of September 11, 2001 (9/11), and have invested trillions of dollars in war and nation-building in Afghanistan and Iraq, in countering al Qaeda and then ISIS,[1] and in caring for returning veterans. Today, after 15 years of inconclusive wars, widespread disenchantment with the country's economic performance, and one of the most polarizing presidential campaigns in recent history, Americans are dissatisfied with their government and divided over their country's role in this unsettled world.

The post–Cold War period is over. While historians may argue about the timing, it has become clear to most foreign-policy practitioners that the world has entered a new era, a complex age of turbulence and opportunity. The challenge of this century is not just to hold the gains of the last century, but to build upon them.

---

[1] The organization's name transliterates from Arabic as al-Dawlah al-Islamiyah fi al-'Iraq wa al-Sham (abbreviated as Da'ish or DAESH). In the West, it is commonly referred to as the Islamic State of Iraq and the Levant (ISIL), the Islamic State of Iraq and Syria, the Islamic State of Iraq and the Sham (both abbreviated as ISIS), or simply as the Islamic State (IS). Arguments abound as to which is the most accurate translation, but here we refer to the group as ISIS.

The first half of this report explores the character of the challenges of what former Secretary of State George P. Shultz has called "a world awash in change." It reviews America's changing ideas of itself and the world beyond as the United States moved from triumph at the end of the Cold War to the current political stalemate and public mood of self-doubt, which has itself become a constraint on engagement abroad. The report also examines the major geopolitical shifts that have occurred since 9/11 in Europe, Asia, and the Middle East, and suggests that a flexible approach to diplomacy will be required to deal with this Rubik's Cube of shifting international alignments.

Existential and Nonexistential Threats

Whether or not "the world is falling apart," the United States is in many ways in an enviable position compared with its rivals. First, the nation faces no certain existential threat. It does face *potential* existential threats from nuclear-armed Russia and China, but without the "we will bury you" animus that put the fear of hot war into the Cold War confrontation with the Soviet Union.[2] Today, these rival states are ambiguously both adversaries and partners—though recently, both appear to be moving more firmly into the adversary camp. Each has made territorial assertions that challenge international conventions of state sovereignty and legal approaches to managing disputes. And each has ramped up spending and made notable military advances that challenge U.S. deterrence.

Russia has emerged as a revanchist power since Vladimir Putin's return to the presidency in 2012 and it now threatens to destabilize Europe through political pressure and subversion. Russia gains a degree of immunity for its aggression by reason of its substantial nuclear arsenal, which it is upgrading. Its "hybrid warfare" tactics, a combination of political and covert subversion and deniable military and cyber operations and information warfare, are more difficult to

---

[2] This remark by Soviet Premier Nikita Khrushchev in 1956 was translated as, "Whether you like it or not, history is on our side. We will bury you." Many scholars argue that what Khrushchev meant was "We will outlast you." See Philip H. Gordon, *Winning the Right War: The Path to Security for America and the World*, New York: Henry Holt, 2007, p. 174.

deal with because of its nuclear capabilities. In the long term, Putin's Russia may well be in decline, due to demographic weakness, overdependence on oil exports, a corrupt and inefficient oligarchical system, and a brain drain. However, declining powers can sometimes be the most dangerous.

China has been modernizing its military for two decades. The growth in its defense spending has almost always outpaced even the spectacular growth of its economy. It has improved its air and missile capabilities, modernized and expanded its navy, and adopted more aggressive postures, particularly in the South China Sea. China is improving its anti-access/area denial capabilities to counter the U.S. ability to project military power throughout East Asia, making it less certain that the United States would win a decisive victory in case of war. As part of the buildup, China is constructing thousands of long-range missiles that could strike targets throughout the region, including major cities such as Tokyo, and even U.S. forces at Andersen Air Force Base in Guam. China has invested in a modern nuclear weapons arsenal that provides it with an increasingly secure strategic deterrent, one that appears to provide Beijing with a key capability that it has long sought and that it was lacking as recently as the mid-1990s: a survivable and thus assured retaliatory capability. While disconcerting, this capability is not automatically bad for the United States because to the extent that it reassures Beijing, it reduces any incentive for a Chinese first strike.

The only unalloyed U.S. adversaries are North Korea and violent jihadist movements, as expressed by ISIS, al Qaeda, and related groups. Threats to U.S. national security may now emanate from strong or authoritarian states that flout international norms, such as Russia or China; brittle states that possess nuclear weapons or the means to develop them, such as North Korea, Pakistan, and Iran; and failing or misgoverned states that cannot or will not enforce the rule of law on their territories, such as Somalia, Syria, or Libya. Threats that once came from "somewhere" can now come from "anywhere," contributing to the meme of a world that is falling apart. Despite the popular perception, a respected index of state fragility found that in aggregate, states became more stable between 1995 and 2013. (The Middle East and Central Africa remained the most troubled regions.) This progress has arguably been purchased with a "peace dividend"[3] that may now have been largely spent. Renewed progress may require a determined partnership between fragile states and those more fortunate. In light of current domestic needs, the focus on fighting terrorism, and the potential costs of international partnerships for dealing with expensive efforts such as mitigating mass migration and climate change, however, it is unclear whether the American people wish to support such a grand bargain between the world's haves and have-nots in light of current domestic needs, the focus on fighting terrorism, and the potential costs of international partnerships for dealing with those expensive efforts.

Familiar, Unfamiliar, and Hybrid Challenges

Although policymakers tend to divide the world by region and devise strategies accordingly, this approach can falter in the face of multidimensional problems that involve the interaction of regional and global trends. A different—and perhaps more useful—way of parsing today's complex set of international challenges is to divide them into categories of familiar and unfamiliar and enduring and emerging, and then tailor expectations, policies, and capabilities to each. Such a framework allows us to separate problems into three categories: familiar challenges that the United States has tackled in the past, with which it can probably cope using existing tools; "hybrid" problems that have both old and unfamiliar characteristics, which will likely require new thinking, tactics, or adaptations; and truly new problems, such as rapid climate change and threats to cybersecurity, which are not well understood or for which appropriate or cost-effective strategies, technologies, or resources are likely lacking. Each type of problem plagues the world today, sometimes in isolation, other times in a specific region, and sometimes coinciding with other problems in time and space.

---

[3] At the end of the Cold War, President George H. W. Bush promised Americans a "peace dividend" in the form of reduced military spending to fight Communism.

Familiarity does not necessarily make problems easier to solve. They may persist because they are expensive, intractable, bloody, or recurring. Often they arise from states that break the international rules and can therefore be addressed by traditional tools of statecraft: diplomacy, defense, deterrence, alliances, economic and military assistance, economic coercion, public diplomacy, subversion, or war. The United States faces a familiar problem from states that are trying to challenge the U.S.-led status quo (China and Russia and—to a lesser extent—North Korea and Iran). Nuclear proliferation is a familiar problem: At the moment, only states have nuclear weapons, and they must be deterred from giving them to other states or to nonstate groups such as terrorist organizations.

Hybrid threats might be defined as challenges that have familiar elements but have evolved in ways that make them even more difficult to deal with. They may develop as a result of a concatenation of familiar problems that might be manageable enough individually, but that create chaos when one compounds another. Hybrid threats will require new approaches, doctrines, training, and—potentially—new international norms for state behavior. For example, Russia's use of "little green men," or irregular soldiers without insignia, in Ukraine poses particular problems for the United States because it represents a clear threat to U.S. interests without rising to the level where it could be considered an attack under existing rules of engagement. Pandemics might be thought of as a hybrid problem. Plagues are ancient, but not the ability of viruses to travel by airplane across continents inside asymptomatic hosts. Hybrid problems may be addressed by what the military calls layered "defense in depth"—the U.S. approach to homeland security since 9/11.

The world also poses transformative challenges that U.S. policymakers have not encountered before and for which they do not currently have satisfactory solutions. By definition, unfamiliar challenges could have far-reaching consequences but their risks are difficult to assess. Moreover, interactions among the increasing number of variables in the current world equation—and hence the compounding or cascading risk—are nearly impossible to anticipate. Policymakers tend to plan and budget for risks that are to some degree foreseen and understood;

efforts are understandably limited when it comes to investigating and preventing low-probability, high-consequence developments for which immediate solutions are deemed unlikely. These emerging and existing problems often do not receive the attention they deserve from senior officials because they are often seen as problems whose solutions—if there are any—lie too far in the future to be worth spending much time thinking about now. In a world of constant technological change, this is a cognitive bias that should be challenged.

Terrorism has become sadly familiar, yet in scale, scope, and ferocity, it is the problem that seems most intractable and largely beyond U.S. ability to control without intolerable sacrifice of the civil liberties on which the "American experiment" is premised. Fear that the government will not be able to protect its citizens against terrorist attacks, including from within, prompts some U.S. voters to favor more-defensive, even isolationist, international policies. Yet it is worth remembering that the problem is not actually new. In the last century, anarchist terrorists conducted numerous bombings in the United States and Europe and successfully assassinated President William McKinley and several crowned heads of Europe, including Archduke Franz Ferdinand, whose death triggered World War I. The United States and Europe have faced major terrorism challenges since the 1970s, but U.S. attempts to understand and counter violent extremism are in their infancy, as are rehabilitation programs being piloted in such countries as Saudi Arabia. Moreover, we would caution against equating the goal of taking back territory seized by ISIS, at which the Russians, the Kurds, the Assad regime, and/or other hard-line Islamist groups may succeed, with "defeating" radical Islam or ending the violence in Syria and Iraq. Containing the threat may be the most sustainable strategy. Some fear that strife among the Sunnis, Shiites, and Kurds may continue in the areas that ISIS controlled and beyond, as Iran tries to create a buffer zone in Shiite-dominated territory in Iraq, the Kurds push for statehood, and radical Sunni fighters from non-ISIS groups fight Shiite domination. Moreover, the assumption that, without territory, ISIS will no longer be able to recruit, inspire, or assist attackers inside Western countries is unproven.

Whatever else is on the next president's agenda, U.S. leadership will be required to address three critical threats to the United States and the planet: nuclear weapons, emerging biological threats, and rapid climate change. Each of these problems has the potential to change or even destroy life as we know it. Despite somber official warnings about the dangers of nuclear weapons and bioterrorism, these grim issues rarely rise to the forefront of public debate. Nevertheless, reducing these supranational threats is a core U.S. interest that must remain a constant government focus.

There is no domestic political consensus about how much the United States should spend to mitigate climate change. Consensus is difficult not only because of ideological divides but also due to multiple great unknowns: the costs of decarbonizing the world economy—that is, of reducing or eliminating net human greenhouse gas emissions; the extent of the damage that will occur if climate change is not controlled; and the timing both of climate change itself and of the human activities to halt it or adapt. This vast range of uncertainty stems both from the novelty of the science—for millions of years, the Earth has not been as warm as it may become, so any estimate of how the climate will behave is an extrapolation of the previously unobserved—and from the difficulty of predicting how different societies and biological systems will respond to the changes. Thus, climate change poses an enormous challenge to policymakers who are required to make decisions with irreversible consequences amid conditions of deep uncertainty.

Leveraging U.S. Strengths, Managing Vulnerabilities

Over the past quarter-century, the advanced industrial economies have been struggling to adapt to three large historic trends: first, the entry of 3.5 billion new people into a globalizing economy; second, rapid technological advances that have created and destroyed whole industries; and third, the shifting demographics of higher-income societies (with the same shift soon to affect middle-income and low-income economies). Each of these transformations is expected to accelerate in the decades to come. Each will require short-term domestic policy adjustments to bend the long-term trend lines in U.S. favor. The United States possesses enormous strengths and competitive advantages that have enabled it to thrive for more than a century in the face of determined adversaries. Chief among them have been the adaptability of the nation and its citizens in confronting daunting challenges, domestic and international; its culture of innovation; and its ability to garner friends and partners across the globe. At the same time, the country also has structural economic weaknesses, deep political and cultural divisions, problems in its labor market, and unprecedented levels of national debt. If not addressed, the political ramifications of these vulnerabilities may constrain the ability of the United States to mount an effective, coherent foreign policy.

There is growing national debate over a wide range of indicators that a segment of American society is failing to thrive, partly because of socioeconomic problems that have not been mitigated by recovery from the Great Recession. The "downward mobility" of some formerly middle-class Americans is more than an economic phenomenon. Americans also now have shorter life expectancy and worse health outcomes than their counterparts in other affluent nations. Mortality rates of white males ages 45 to 54 have increased for the first time,[4] as has the suicide rate.[5] A Federal Reserve survey found that 31 percent of Americans said they were "struggling to get by" or "just getting by" financially.[6] This failure to thrive among nearly one-third of the population raises the question of how to improve job prospects, economic well-being, and financial security, without hurting the incentive systems that keep the U.S. economy strong. Whatever future administrations' policies toward the "1 percent"

---

[4] Anne Case and Angus Deaton, "Rising Morbidity and Mortality in Midlife Among White Non-Hispanic Americans in the 21st Century," *Proceedings of the National Academy of Sciences*, Vol. 112, No. 49, December 8, 2015.

[5] Erin M. Sullivan et al., "Suicide Among Adults Aged 35–64 Years—United States, 1999–2010," Centers for Disease Control and Prevention, *Morbidity and Mortality Weekly Report*, Vol. 62, No. 17, May 3, 2013, pp. 321–325.

[6] Consumer and Community Development Research Section of the Federal Reserve Board's Division of Consumer and Community Affairs, *Report on the Economic Well-Being of U.S. Households in 2015*, Washington, D.C.: Board of Governors of the Federal Reserve System, May 2016.

of richest Americans, its policies toward the "31 percent" of lower-income Americans may prove most crucial in restoring political unity, as well as the U.S. position in the world.

At the same time, debt, jobs, dislocation, and demographics are problems for many advanced economies, and the United States is better positioned than any of its rivals to address them. Its first advantage is the power of agency. The reforms needed to sustain its economic vitality are within America's power to enact. This is not the case for some other countries, which are more subject to external pressures—the strength of the global economy, the changes in market demand and supply conditions, and their levels of debt, among other factors. Although segments of the American public are increasingly skeptical of globalization, the evidence that the United States benefits from the global economy is compelling. The United States also possesses strengths and capabilities that position it well to thrive in the 21st-century economy, and will likely keep it the world's largest economy at least for several decades to come. Specific policies should be considered to spur rapid adaptation to change; improve education, particularly the number of college graduates; improve recruitment and retention of global talent; manage population aging so it does not depress growth; and reverse the decline in spending on federal research and development. Whatever the specific policy chosen, maintaining a climate that attracts and promotes entrepreneurship and investment, along with building an economy that delivers on the promise of the American dream for a larger share of the population, should be part of an "opportunity agenda" for any administration. Success at home begets strength abroad; it strengthens any American president's power of persuasion. International strategy will be most successful when it reflects the recognition that the U.S. power to attract is greater than its power to compel.

Anticipation, Deterrence, and Resilience

Globalization, the increasing numbers of world actors and variables, and the complex and concurrent interactions among them, have increased uncertainty, defined as a state in which "information is too imprecise to be summarized by probabilities."[7] Uncertainty is an intractable problem for a nation that spends more than $600 billion per year to defend against security threats, actual and anticipated.

How can the U.S. government improve its ability to make better decisions in an uncertain and perhaps incoherent environment? Three policymaking approaches are particularly relevant in highly uncertain environments and can be improved: anticipation, deterrence, and resilience. We must anticipate what we can, and act on this foresight; rethink how to deter those threats we can anticipate; and build resilience to withstand and rebound from those attacks, surprises, or calamities that we cannot anticipate or deter. The United States needs a comprehensive strategy that includes investments in anticipation (to manage risk, allocate resources wisely, avoid or minimize shocks, prevent conflicts, and prepare for foreseeable outcomes), deterrence (to prevent politico-military competition from escalating into war), and resilience (to withstand shocks that were not prevented or perhaps preventable). The United States could also benefit from working on a number of elements of the anticipation equation, including understanding surprise; improving intelligence collection, analysis, and absorption; creating bureaucratic mechanisms to combat cognitive bias and integrate thinking about low-probability, high-consequence problems into the policymaking process; distinguishing between long-term problems and low-urgency problems; and using decision aids, including teaming up humans and computers.

Many of the assumptions that animated U.S. foreign policy in the Cold War period no longer apply. The instruments of warfare and the character of economic competition have evolved. So have the strategic orientations of Russia and China. These changes require a reexamination, and in a fundamental sense, a (re)learning of what it means to deter. One of the foremost, if not most pressing, reasons for such a

---

[7] Ian Goldin and Mike Mariathasan, *The Butterfly Defect: How Globalization Creates Systemic Risks, and What to Do About It*, Princeton, N.J.: Princeton University Press, May 11, 2014, p. 25. In the early 20th century, the economist John Maynard Keynes and the statistician Frank Knight were among the first to formally distinguish between *risk*, which can be quantified, and *uncertainty*, which cannot. In recent years, the term *deep uncertainty* has been used for information that is too imprecise to be summarized with confidence or expressed by quantified probabilities.

reexamination arises in connection with questions—actual and potential—over the credibility of U.S. plans to defend allies against attack, including threats to employ nuclear weapons, if necessary. This is not a new concern; rather it is one that has been thrust back onto the agenda of decisionmakers over the past several years as it has become apparent that China and Russia are developing forces, doctrine, and weapons designed to achieve military objectives against nearby states before U.S. and allied forces can intervene in sufficient numbers to make a difference.[8] U.S. strategic thinkers and defense analysts share the allies' concerns that emboldened Chinese and Russian leaders might risk testing U.S. resolve by seizing territory and waters claimed by a U.S. ally in East Asia or recapturing one or more of the Baltic states—in short, to present the United States with a *fait accompli*.

Against *fait accompli* strategies, nuclear forces—which intrinsically promise to punish the enemy for undertaking a specific act—may lack needed credibility in the eyes of many adversaries for communicating the willingness to defeat adversary aims should deterrence fail. Moreover, the sheer geographic proximity of the strategic objectives that Beijing and Moscow so covet (e.g., the Spratly Islands and Paracel Islands in the South China Sea and the Baltic states), combined with vastly improved Chinese and Russian military capabilities, make *fait accompli* strategies comparatively easier to accomplish than at any time since the end of the Cold War.

Therefore, to meet the multiplicity of actors and the multiplicity of challenges the United States is likely to face in the decade ahead, the nation's immediate focus must be on (1) emphasizing conventional deterrence—i.e., conventional capabilities relative to nuclear deterrence capabilities—(2) preparing to deter across multiple domains, and (3) tailoring deterrent capabilities—conventional and nuclear alike—and messages to influence different potential adversaries.

From the standpoint of national strategy, active steps to improve resilience in many aspects of government planning and preparedness will

---

[8] For instance, China's strategy of "winning informationized local wars" envisions localized, short-duration, and high-intensity conflicts. See State Council Information Office of the People's Republic of China, *China's Military Strategy*, Beijing, May 2015.

also be particularly important in the coming decades, given the increasing recognition that the more globalized and complex our societies become, the more exposed they will be to systemic global risk. Of particular concern is the lack of preparedness for next-generation cascading disasters—so-called "global network disasters" or "network of networks" disasters. Preparation and preventive action may be urgent even when the perception of risk is not. Humans tend to value future benefits less than immediate ones, but this is a cognitive error. Assuming that more-frequent or more-severe disasters or shocks could be the new norm, key elements of resilience that can be encouraged by the federal government in its domestic and international planning include preparing for a long recovery period, and recognizing that another disaster may strike while a community is still recovering from the first; establishing distributed, networked systems of communication and response that can offset the destruction of centralized governmental systems (for example, from terrorism or cyberattacks); encouraging a more collaborative style of leadership and management; ensuring that management systems are not so "lean and mean" (or dependent on single supply chains for vital relief supplies) that they cannot ramp up in times of disaster; and redesigning policies, programs, infrastructure, and virtual networks to strengthen adaptability, recovery, and resilience capabilities.

Strategic Choices

The second half of this report examines the nature of strategy and proposes three plausible alternative strategic concepts reflective of the contemporary political debate. A presidential formulation of an overarching statement of national purpose and priorities for engaging the world—better put, a strategic concept or orientation—has value in defining broad national goals and priorities and providing the political rationale for domestic and allied support. It gives direction to diplomacy and defenses, promotes bureaucratic coordination, and enables more-efficient use of resources. Despite the limitations of strategizing and policy planning in a turbulent world—to paraphrase President

Dwight Eisenhower—while plans in real-world application may turn out to be worthless, planning is everything.

This report defines the core U.S. interests that are—or should be—common to any future administration. The primary national interest of the United States is to advance the security, freedom, and prosperity of the American people, so that they may enjoy, as the Founding Fathers put it, "life, liberty, and the pursuit of happiness." Over time, national leaders have enunciated a large number of subsidiary interests, sub-subsidiary interests, and doctrines for how to achieve them. True "interests" (which must be defended) should not be confused with "objectives" (which evolve over time and will be pursued by most administrations wherever possible but are not essential to the survival of the country). On closer inspection, many so-called "interests" turn out to be mutable. However, this report lists nine core objectives that the next administration will need to achieve, at a minimum, in order to defend the national interest:

- Protect the U.S. population and territory from nuclear attack and other forms of irreversible harm.
- Increase U.S. economic growth and broad-based prosperity, and retain U.S. global economic leadership of the liberal, rules-based economic order.
- Prevent hostile domination of Europe, East Asia, and the Middle East.
- Undercut the ideologies of violent Salafi-jihadism and reduce the capabilities of groups espousing it.
- Make good on U.S. commitments to NATO, the Asian allies, Israel, and others, while persuading allies to do more for their self-defense.
- Limit the number of states with nuclear weapons. Combat proliferation of nuclear and biological weapons and leakage of sensitive advanced technologies.
- Advocate respect for universal human rights and protection for freedom-seeking people.
- Retain control over U.S. borders to prevent illegal migration or terrorist infiltration.

- Reduce the risk of climate change to the United States.

There is no single way to satisfy these interests or objectives; thus, we consider three different strategic concepts or orientations:

- "Come Home America:" international restraint and domestic renewal
- "The Indispensable Nation:" America as promoter of world order
- "Agile America:" adapt and compete in a changing world.

It is important to consider not only the foreign policy outcomes Americans desire, but also imperfect ones they may well have to live with. Therefore, "Come Home America" adopts a narrow definition of U.S. interests, sets modest global objectives, and accepts minimally acceptable outcomes, in exchange for limiting U.S. exposure to a more unstable world. This orientation would accept that some—indeed many—international problems are beyond U.S. ability to control. Therefore, it would focus on what is within U.S. control, prioritizing domestic growth and improved international competitiveness, while attempting to limit military and political commitments abroad. It would reject "liberal hegemony," but stay engaged in the world for the purpose of advancing U.S. economic interests abroad. Instead of military "boots on the ground," it would prefer to put American-made boots on store shelves around the world. It assumes that both the national security and discretionary domestic budgets will continue to be constrained by debt, slow growth, and political polarization. It would narrow the definition of "core" U.S. national security interests, aiming to *spend* less and *do* less abroad.

"The Indispensable Nation" embraces a more traditional view of the United States as the leader and enforcer of global order. It sees America's role as the chief promoter of the liberal international order and aspires to more maximalist goals. In this worldview, the world without U.S. leadership is a Hobbesian place. Active leadership and engagement, whatever their flaws, are deemed vastly preferable to chaos. The steeper costs are outweighed by the gains for U.S. citizens and the world. In this view, the devastating civil war in Syria, the rise of ISIS, the migration crisis, Russia's military involvement in Syria, and rising instability across the Middle East in the wake of the failed "Arab Spring" are the second- and third-order consequences of premature U.S. disengagement from Iraq and Libya. In this view, only the United States, with all its warts, has the resources, prestige, and expertise to lead, and the United States is at its best when advancing and defending a benign liberal international world. Failure to invest in global security in this moment of turbulence and change will diminish U.S. power, security, and prosperity—and no other power is likely to shape the world to such benign ends.

"Agile America" would keep the United States engaged in the world, but primarily for the purposes of advancing the economic well-being of the United States and its friends. It accepts globalization as inevitable and even desirable, and would use U.S. foreign policy and economic statecraft to help make it work better for the American people. Its goal is to manage change in such a way as to benefit all Americans. In this view, it will be nearly impossible, and probably undesirable, to try to limit U.S. exposure to an unstable world, as retreat would diminish security as well as economic and political opportunities for the United States. At the same time, the cost of remaining "the indispensable nation" will be unaffordable. The United States would remain engaged in the world, but with foreign policies prioritizing the economic well-being of the United States and limiting what appear to be expensive security commitments, particularly with partners able to do much more on their own behalf. Like the first option, it would attempt to set limits on U.S. global security obligations—but instead of curtailing them, it would attempt to adapt them in ways that reflect the nation's changing geo-economic interests.

The report identifies the core beliefs and assumptions behind each strategic concept, and interrelates broad policy objectives, degrees of international activism, general budgetary implications, constraints, and risks. The policies chosen to implement them can be fluid, often dictated not by ideological or strategic considerations, but by fast-changing circumstances, diplomatic necessities, financial constraints, and above all, judgments about what is likely to work.

We do not endorse or recommend any of these three strategic concepts and assume that the next administration will mix and match policies as it deems fit.

Finally, at a time of deep partisan division, the report offers guidance not only on which paths might be the most promising, but also how to choose among various strategies. The report offers six criteria that might guide decisionmaking, although no single consideration will be determinative. These are: scrutinize assumptions, seize opportunities, uphold commitments, play both the long and the short games, align U.S. interests and values, and limit regret.

The report concludes that domestic political dysfunction is the greatest obstacle to effective U.S. global leadership. A coherent international strategy will be difficult to pursue without a greater degree of domestic political consensus; a "grand strategy" is simply impossible. Building domestic political consensus on the U.S. role in world affairs would require bipartisan agreement to reconcile the competing demands of domestic needs with the requirements of effective international engagement. U.S. leaders would also need to break the political deadlock between funding requirements for the desired degree of international engagement and commensurate levels of taxation. On the basis of such reconciliation, U.S. diplomacy, defenses, and alliance relationships could be adequately resourced to support the chosen strategy. The government could develop coherent plans to promote economic growth, national security, and international engagement, as required in a world where domestic well-being is inextricably linked to global developments.

Without such political consensus, the United States will face heightened risks.

Acknowledgments

RAND Corporation President and Chief Executive Officer Michael D. Rich conceived the Strategic Rethink project and has supported it through the research phase and the production of the six associated volumes. The authors particularly wish to thank Rita Hauser and the Hauser Foundation for supporting and sustaining our efforts.

As with most RAND projects, Strategic Rethink incorporates the thinking and work of a range of RAND staff and contributions by numerous individuals in the policy community. The authors of the previous volumes in this series have also contributed generously to this one. We thank Hans Binnendijk, Michael S. Chase, Ryan Henry, Seth G. Jones, F. Stephen Larrabee, Andrew Liepman, Jeffrey Martini, David Ochmanek, Charles P. Ries, James T. Quinlivan, and the late Edward L. Warner.

Special appreciation is due to Ambassador James Dobbins, the primary author of the first volume in the series, *Choices for America in a Turbulent World*, for his leadership, insights, and substantive contributions. We also thank Bruce Jones, Michael J. Mazarr, and Andrew Parasiliti for their insightful and constructive comments on the manuscript.

We are grateful to the many other friends and colleagues who have contributed to our thinking about the many and complex issues related to international relations and U.S. engagement in the world: Lillian Ablon, Lina Andrews, Robert Bozick, Nicholas Burger, Christopher S. Chivvis, William Courtney, Kim Cragin, Chester A. Crocker, Daniel M. Gerstein, David C. Gompert, Adam R. Grissom, Jennifer Kavanagh, Krishna B. Kumar, Charles Hill, Brian Michael Jenkins, the late Andrew Kohut,

Martin C. Libicki, Melinda Moore, Thomas Pickering, David Rothkopf, K. Jack Riley, Richard Silberglitt, Simon Sinek, and Ali Wyne.

We also want to thank Sarah Harting, Stacy Fitzsimmons, Maxine Morris, Cecile St. Julien, Jamie Greenberg, and Kathleen Cutsforth for their research and administrative support, and Arwen Bicknell for her expert editing.

CHAPTER ONE

Introduction

What is America's role in the world of the 21st century? And is there a strategic concept—a "grand strategy"—that would give direction to our diplomacy, defense, and economic engagement with the world?

These are the orienting questions of the RAND Strategic Rethink project, which has assessed the diverse challenges—and opportunities—for the United States abroad in this time of transformation and turbulence.

For more than seven decades, the United States has assumed leadership in building a liberal international order.[1] From 1947 until the end of the Cold War in 1991, the United States confronted threats from the totalitarian, nuclear-armed Soviet Union and from other authoritarian states. After the Cold War, the United States organized collective responses to challenges to the international order posed by less-powerful aggressors, including tyrants who committed grave violations against their neighbors or their own populations. Americans united after the terrorist attacks of September 11, 2001 (9/11), and have invested trillions of dollars in war and nation-building in Afghanistan and Iraq, in countering al Qaeda and then ISIS,[2] and in caring for returning veterans. Today, after 15 years of war, wide-

---

[1] See G. John Ikenberry, *Liberal Leviathan: The Origins, Crisis, and Transformation of the American World Order*, Princeton, N.J.: Princeton University Press, 2011; and Robert Kagan, *The World America Made*, New York: Knopf, 2012.

[2] The organization's name transliterates from Arabic as al-Dawlah al-Islamiyah fi al-'Iraq wa al-Sham (abbreviated as Da'ish or DAESH). In the West, it is commonly referred to as the Islamic State of Iraq and the Levant (ISIL), the Islamic State of Iraq and Syria, the Islamic State

spread disenchantment with the country's economic performance, and a deeply polarizing presidential election, Americans remain as divided as ever about the role they envision for their country in this unsettled world.

The public mood shifted from a sense of triumphalism following the collapse of the Soviet Union to frustration with the costs and results of large-scale U.S. nation-building efforts, skepticism about involvement in local conflicts and crises of seemingly uncertain relevance to U.S. interests (in Somalia, the Balkans, Rwanda, Haiti, Afghanistan, Iraq, and now Syria), and domestic political stalemate. The spreading threat from Islamist jihadism and the large number of terrorist attacks by ISIS and its sympathizers around the world have renewed questions about why violent Salafi-jihadism still has not been contained, much less defeated.

More broadly, Americans are debating the ability of any non-Muslim power to counter the ideologies of radical Islam, the Shiite-Sunni rift, and other forms of sectarian violence that are ravaging the Islamic world. Can the West intervene militarily without strengthening the ranks of violent extremists who seek to provoke an apocalyptic conflict between the Islamic and Christian worlds? Can the United States conduct the intelligence and security activities necessary to prevent terrorist attacks without undermining civil liberties at home?

Some question why the United States must so frequently resort to the use of force,[3] or bear the costs of stabilizing failing or oppressive governments, some of which are U.S. partners in countering terrorism.[4] Others fear that a more assertive China and a revanchist Russia are leading the world back to confrontations that were supposed to have ended with the Cold War. Many see no choice but a firm U.S. response. There is broad agreement that allies in Europe and Asia should improve their military capabilities, yet resentment

---

of Iraq and the Sham (both abbreviated as ISIS), or simply as the Islamic State (IS). Arguments abound as to which is the most accurate translation, but here we refer to the group as ISIS.

[3] See Stephen M. Walt, "Is America Addicted to War?" *Foreign Policy*, April 4, 2011.

[4] See Matt Schiavenza, "Why the U.S. Is Stuck with Saudi Arabia: Even the Shale Oil Revolution Can't Wean Washington off its Despotic Middle Eastern Ally," *The Atlantic*, January 2015.

that some allies have prospered under the U.S. security umbrella while underinvesting in their own defense.[5] In a 2016 Gallup poll, 67 percent of those surveyed said it was important that the United States be the number one military power in the world—though only 49 percent believed it to be so.[6]

For six decades, a majority of the American public supported active U.S. engagement in the world, with a sharp dip during and after the Vietnam War, as shown in Figure 1.1.[7] The 9/11 attacks boosted public support, but the Iraq war and its aftermath triggered a sharp reversal. In 2013, more Americans than ever agreed with the statement that the United States should "mind its own business internationally." Yet recent polling suggests that public opinion may be shifting again toward a more activist role, in response to the ISIS threat.[8] The U.S. public continues to rank the economy and terrorism as the two top national priorities. It is unclear whether growing international threats combined with full recovery from the Great Recession will translate

---

[5] Public irritation over allied "free-riding" is bipartisan, and not restricted to the United States. See, for example, James Joyner, "Europe's Free Ride on the American-Defense Gravy Train," National Review Online, July 13, 2014. An Afghan blogger warned that "the U.S. and its NATO allies spent much blood and treasure in Afghanistan, but China will reap the economic benefits on a free ride" (Abdullah Sharif, "China's Ambitions in Asia," *World Post*, November 13, 2015). Former Rep. Barney Frank complained in an article that "Paradoxically, we continue to punish Germany and Japan for the horrors they inflicted on the world decades ago by, in part, freeing their resources to be spent on their domestic well-being" (Barney Frank, "It's Time to Rearm Germany and Japan," *Politico*, October 21, 2015).

[6] "U.S. Position in the World," Gallup, web page, February 3–7, 2016.

[7] Gallup has documented high levels of distrust in the government and rising dissatisfaction with the U.S. role in the world. "U.S. Position in the World," 2016.

[8] For example, in CNN/ORC polling, support for sending U.S. ground troops to fight ISIS in Iraq or Syria rose from 38 percent in September 2014 to 48 percent in May 2016, with 48 percent still opposed. CNN/ORC International, "CNN/ORC International Poll," December 6, 2015; CNN/ORC International, "CNN/ORC International Poll," May 5, 2016. However, the April 2016 Pew Poll of 1,000 respondents found 46 percent in favor of sending ground troops and 50 percent opposed. Respondents were sharply divided along partisan lines, with 68 percent of Republicans favoring the use of ground troops and 65 percent of Democrats opposed. Pew Research Center, *Public Uncertain, Divided Over America's Place in the World*, Washington, D.C., May 5, 2016b.

Figure 1.1
Fluctuating Support for U.S. International Engagement

The U.S. should mind its own business internationally and let other countries get along the best they can on their own.

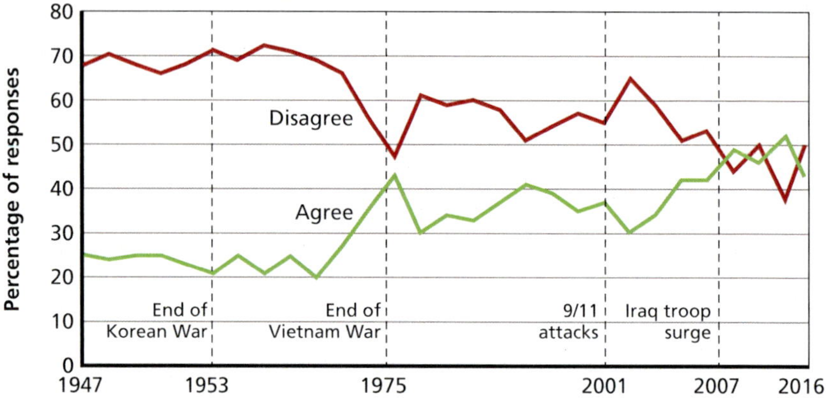

SOURCES: Dina Smeltz, *Foreign Policy in the New Millennium: Results of the 2012 Chicago Council Survey of American Public Opinion and U.S. Foreign Policy*, Chicago, Ill.: Chicago Council on Global Affairs, 2012; Pew Research Center, 2016b.
RAND *RR1631-1.1*

into renewed public willingness to support a more activist U.S. foreign policy.[9]

Polling data suggest that the public mood reflects more than a lack of confidence in the government's ability to eliminate terrorism and stabilize the Middle East;[10] it also reflects a loss of political consensus about America's role in the world and, more generally, a high level

[9] In a reversal of long-time trends, the Chicago Council Survey found that Democrats are now more likely than Republicans to support an active U.S. role in world affairs. In 1974, for example, 72 percent of Republicans agreed the United States should "take an active part in world affairs," in 2006, 77 percent did. By 2016, that number had fallen to 64 percent. Democratic support for an active role was 68 percent in 1974, 65 percent in 2006, and 70 percent in 2016. Dina Smeltz et al., "America in the Age of Uncertainty," Chicago, Ill.: Chicago Council on Global Affairs, 2016, based on a survey of 2,061 adults from June 10 to June 27, 2016.

[10] "U.S. Position in the World," 2016. Gallup has documented high levels of distrust in the government and rising dissatisfaction with the U.S. role in the world. The Pew Research Center documented similar results in a late 2015 poll that found that only 19 percent of Americans say they can trust the government in Washington most of the time—down from

of distrust of the government and its ability to manage today's international challenges. (The United States is not alone in this trend, as seen in the June 2016 "Brexit" referendum in the United Kingdom, and the significant anti-Brussels sentiment in other European Union [EU] member countries.) The cleavages and uncertainty in public opinion pose a major challenge to forging a coherent international strategy, whether it is activist, restrained, or a retooled approach to globalization as a way to jump-start U.S. prosperity. This is the divided and fluid public mood that will confront the next administration—and likely its successors.

The lead volume in the Strategic Rethink series, by Ambassador James Dobbins, laid out choices for the next administration in key regions and issues, and uses an operational definition to identify the ways and means applied to achieving specific and immediate policy objectives. In this concluding volume, we conceptualize three alternative "grand" strategic orientations that reflect alternative views of America's circumstances and how to advance the country's interests in a turbulent world. We then offer thoughts about how to choose among these strategies, identifying criteria that may be helpful to officials facing difficult policy decisions.

The first half of this assessment explores in some detail the character of the challenges of what former Secretary of State George P. Shultz has called "a world awash in change."[11] Chapter Two reviews America's changing ideas of itself and the world beyond as the United States moved from triumph at the end of the Cold War to the current political stalemate and public mood of self-doubt, which has itself become a constraint on engagement abroad. Chapter Three analyzes the nature of the world today, untangling the strands of change and turmoil that cause many to wonder whether the world is in fact "falling apart." These include geopolitical realignments, as well as technological, social, environmental, and economic developments that any

a record high of 77 percent in 1964. Pew Research Center, *Beyond Distrust: How Americans View Their Government*, Washington, D.C., November 23, 2015a.

[11] Quoted in David Horovitz, "George Shultz: In a World Awash in Change, Follow the Reagan Playbook," *Times of Israel*, February 21, 2016.

national strategy must account for. It discusses the revival of familiar challenges, such as assertive major-power competitors; hybrid challenges, such as Salafi-jihadist terrorism or the "little green men" who have done much of the fighting in Ukraine; and unfamiliar problems for which we do not now have satisfactory solutions, such as abrupt climate change or pandemics. It identifies three potentially existential threats to the United States in the 21st century: nuclear weapons, bioterrorism, and climate change.

Since successful strategy involves leveraging advantages as well as minimizing vulnerabilities, Chapter Four turns to the chief sources of U.S. strength and vulnerability. It discusses the domestic issues that are particularly relevant to the country's ability to sustain its international engagement: jobs, innovation, and demographic challenges. Chapter Five discusses three specific policymaking approaches that the United States could cultivate to prepare for this uncertain world: anticipation, deterrence, and resilience.

The second half of the report analyzes alternatives for the path forward. It begins with an overview of the ongoing efforts, in academia and the think tank world, to develop a "grand strategy" to give greater coherence to U.S. defense, diplomacy, and economic engagement abroad. It concludes that the United States needs a strategic concept—a basic set of operating principles and assumptions—to guide the innumerable tactical decisions of international engagement and to develop public understanding, political backing, and allies' support. It then offers a definition of core U.S. interests that are—or should be—common to any future administration, and objectives to be pursued. It then suggests three possible strategic orientations, each reflected in varying ways in the contemporary national dialogue:

- "Come Home America:" domestic renewal and international restraint
- "The Indispensable Nation:" America as promoter of world order
- "Agile America:" adapt and compete in a changing world.

It is important to consider not only the foreign policy outcomes that Americans desire but also imperfect ones they may well have to live with. Therefore, we begin with an assessment of these three strategic orientations by considering a posture of international restraint and domestic renewal. This option, which we call "Come Home America" adopts a narrow definition of U.S. interests, sets modest global objectives, and accepts minimally acceptable outcomes, in exchange for limiting America's exposure to a more unstable world and freeing up resources for domestic renewal. The second strategic option embraces the traditional view of the United States as an "indispensable nation."[12] It sees America's role as the chief promoter of the liberal international order and aspires to more maximalist goals. The third strategic concept would keep the United States fully engaged in the world, but primarily for the purposes of advancing the economic well-being of the United States and its friends. Like the first option, it would attempt to set limits on U.S. global security obligations. Readers who remember previous political campaigns might think of this as "It's the (Global) Economy, Stupid."[13] It accepts globalization of the economy as inevitable and even desirable, and uses U.S. foreign policy to help make it work better for the American people. Each alternative is developed as a plausible strategic concept in the context of the contemporary American debate. We do not endorse or recommend any of them. We assume that policymakers will mix and match policies as circumstances demand.

For each strategic orientation, we explore the core political beliefs and underlying assumptions about how the world works. The saying in government that "he who controls the assumptions controls the policy" remains true even as an administration revisits and revises its thinking over the course of its term. After analyzing the assumptions, we correlate options for global activism with budgetary constraints and levels of risk—recognizing that risk perceptions are subjective and also that

---

[12] This term was coined by Madeleine Albright: "If we have to use force, it is because we are America; we are the indispensable nation. We stand tall and we see further than other countries into the future, and we see the danger here to all of us." Madeleine Albright, *Today*, NBC, February 19, 1998.

[13] This was the campaign slogan of James Carville, Bill Clinton's campaign manager, during the 1992 presidential race. Carville held that many voters who did not believe that the 1990 recession was actually over would vote against the party in power. See Robert J. Samuelson, "It's Still the Economy, Stupid," *Washington Post*, February 3, 2016.

neither increased expenditure nor international retrenchment automatically translate into reduced risk to U.S. interests.

These illustrative strategic orientations are not meant to be caricatures of views that lie along the traditional American foreign policy spectrum. "International restraint" does not mean across-the-board retrenchment or isolationism, though it would certainly entail choices about where and when to commit resources. The President might actively engage in one region, relying on military pressure to achieve some ends, while prioritizing diplomatic and economic tools in another region, and declining to be drawn into a conflict in another area. Strategic selectivity—avoiding committing the United States to lead on every issue and bear every burden—is a necessity, not a weakness.

Likewise, the second, activist orientation implies U.S. leadership but not interventionism or automatic support for the use of U.S. military forces, especially large-scale use of ground forces. Nor does it assume that to remain "indispensable," the United States must engage everywhere at all times to defend humanitarian norms or enforce international law. A President who exercised restraint over U.S. involvement in the Middle East might launch an ambitious effort to develop an Asian security architecture, shore up NATO security, or reduce global greenhouse gas emissions.

The third option, "Agile America," does not merely seek a "moderate" middle ground between restraint and activism. Rather, it would keep the United States fully engaged in the world while prioritizing pursuit of its economic interests. Recognizing that globalization of the economy is unstoppable and ultimately beneficial, it aims to retool U.S. policy to help the country thrive amid the geopolitical and geo-economic shifts already under way and those to come.

All three strategic concepts are predicated on at least three overarching assumptions that have surfaced over the course of the Strategic Rethink project, and that deserve increased attention from the U.S. policymaking community.

First, all three orientations are based on the assumption that the United States does not play the global game alone. An enduring source of America's strength is its alliances and international partnerships. As

Bruce Jones points out, the United States has the most desirable friends in the world, and many more of them than its rivals can claim.[14] Moreover, its allies and friends are mainly consolidated or emerging democracies with market-based economic systems and a commitment to universal human rights. Their governance is based on the rule of law, they support international norms, and they share many common interests (including the desire to gain more influence to shape the existing international order to their benefit, but not to upend it).

Second, we conclude that the liberal international order built over the past seven decades, despite its flaws, has well served the United States and much of the world.[15] The liberal order has resulted in fewer interstate conflicts than in the previous hundred years; the unprecedented creation of wealth and a historic decline in poverty;[16] a doubling of literacy rates;[17] a dramatic improvement in human rights—including women's rights—and individual freedoms in large areas of the world; and an explosion of knowledge, invention, and creativity. Freedom of expression—even the "right" to poke fun at powerful leaders—is now enjoyed by citizens of at least 86 countries.[18] To be sure, the benefits of the last seven decades have been unevenly distributed, leaving a toxic legacy of misery, resentment, and political strife in the places left behind. Precisely for this reason, the U.S. national interest lies in preserving and extending the liberal democratic order where possible, and ensuring that its benefits are more widely shared in the decades to come. The specific strategies and tactics that will accomplish this goal are open to debate.

---

[14] Bruce Jones, *Still Ours to Lead: America, Rising Powers, and the Tension Between Rivalry and Restraint*, Washington, D.C.: Brookings Institution Press, 2014, p. 201.

[15] Hal Brands, *American Grand Strategy and the Liberal Order: Continuity, Change, and Options for the Future*, Santa Monica, Calif., RAND Corporation, PE-209-OSD, 2016.

[16] For an overview of the progress of the last 15 years, see United Nations (UN), *Millennium Development Goals Report*, New York: Department of Economic and Social Affairs, 2015a.

[17] By 2010, 83 percent of the world was literate, up from 42 percent in 1940. Max Roser and Esteban Ortiz-Ospina, "Literacy," OurWorldInData.org, website, 2016.

[18] Despite setbacks in the past decade, Freedom House rated 86 countries as "free," 59 as "partly free," and 50 as "not free." Freedom House, "Freedom in the World: Country Ratings 1973–2016," 2016b.

While Americans will likely continue to view their nation as an exceptional force for good in the world, we can expect the United States to find its preferences challenged by increasingly strong rivals. Defending the liberal order does not mean that the United States will be able to advance its values everywhere at all times. Yet we assume that it must continue to work with its liberal democratic allies to advance humanitarian goals, human rights, and individual dignity wherever it can; no other nation will. In the aftermath of the "Arab Winter," it may choose not to push for transitions to democracy in the most troubled areas of the world, where the precursors of civil society are lacking. It may opt not to put troops on the ground to live up to the new international norm of the "responsibility to protect" (R2P, discussed in Chapter Two). Yet, especially in turbulent times, the United States can increase its efforts to support and strengthen the other 126 established and emerging democracies. It is on their success that the liberal international order ultimately depends.[19]

Third, it is safe to assume that uncertainty will continue, and perhaps even intensify; it is a constant in human affairs. This does not mean that the world is "falling apart," as some fear; but for Americans who have lived through sunnier times, it is turbulent, disorienting, and unsettling. As former national security adviser Brent Scowcroft put it, "This world is not as dangerous as that during the Cold War, but it is *much* more complicated."[20]

In such a complex and dynamic environment, strategic surprises may or may not occur more frequently than in decades past—this is unknowable. But we can safely assume that surprises will happen. Futurists Peter Schwartz and Doug Randall define *strategic surprises* as "events that, if they occur, would make a big difference to the future,

---

[19] Ted Piccone argues that the success of Brazil, South Africa, India, Turkey, and Indonesia may determine the trajectory of the liberal order. To this, however, might be added several of the increasingly fragile democracies of Eastern Europe, as well as the Western European nations in political turmoil as a result of the migration crisis. Ted Piccone, *Five Rising Democracies and the Fate of the Liberal Democratic Order*, Washington, D.C.: Brookings Institution Press, 2016.

[20] As quoted in David Rothkopf, *National Insecurity: American Leadership in an Age of Fear*, New York: Public Affairs, 2014, p. 350.

force decision makers to challenge their own assumptions of how the world works, and require hard choices today."[21] For example, 9/11, the 2007–2008 financial crisis, the Arab Spring, and the Russian absorption of Crimea were surprises that ultimately required significant shifts in assumptions and policy.

In hindsight, some of the strategic surprises to come may turn out to have been possible to anticipate, whereas others may be so-called "black swans,"[22] events that were so unlikely that even the best intelligence probably would have failed to consider them. The anatomy of the anticipation of problems is dissected in more detail in Chapter Five. We conclude that much more attention to anticipation is necessary.

The sheer number of variables in the international equation—many more capable states and nonstate actors, the global movement of people and capital, and accelerating technological developments—increase the possibility of compounding or cascading risks. At the same time, the scale of that disruption is expected to increase as a sheer function of projected global population growth, the concentration of people in megacities, and climate change.[23] Without more-effective interventions, the human toll from future conflicts, natural disasters, pandemics, and rapid climate change can be expected to grow much worse. Policymakers are faced with myriad competing challenges—familiar and unfamiliar, likely and improbable, politically expedient and politically dangerous—with limited resources to allocate. How can they plan a strategy in the midst of such uncertainty?

---

[21] As quoted in Nathan Freier, *Known Unknowns: Unconventional 'Strategic Shocks' in Defense Strategy Development*, Carlisle, Pa.: U.S. Army War College, November 2008, p. 5.

[22] Nassim Nicholas Taleb, who coined the phrase, observes that "our tools for forecasting and risk measurement cannot begin to capture black swans." Nassim Nicholas Taleb, "Learning to Love Volatility," *Wall Street Journal*, November 16, 2012.

[23] The United Nations estimates a population increase of more than 1 billion between 2015 and 2030, increasing from 7.3 billion in 2015 to 8.5 billion in 2030 and up to 11.2 billion in 2100. Most of the growth will occur in the developing world, where population density is projected to rise from 25.5 people per square kilometer in 2015 to 86.2 people in 2100. UN, *World Population Prospects, 2015 Revision*, New York: Department of Economic and Social Affairs, 2015b.

This volume argues that they must develop strategies that not only advance U.S. interests but also hedge against failure. Furthermore, any national strategy for this turbulent age must strengthen the U.S. capacity to anticipate, prevent, or withstand the inevitable surprises. Chapter Five discusses three specific policymaking approaches that the United States could cultivate to prepare for this uncertain world: anticipation, deterrence, and resilience. Working on improving anticipation can help federal, state, and local governments consider risk along a better-defined spectrum, and help policymakers prioritize their efforts. Working on deterrence can better clarify U.S. interests and stakes, demonstrate those stakes to would-be aggressors, and thereby avoid war. Working on resilience can help the nation withstand shocks and "build back better" physical and social infrastructure than what may be destroyed by natural or manmade disasters. No matter what path the United States follows, it will require a sophisticated understanding of deterrence and resilience to form the bedrock of national security. We also suggest that the next President and Congress update the mechanisms of national security planning and decisionmaking to foster more agile and effective governance.[24]

The concluding chapter challenges the reader to think more deeply not only about *which* strategic concept and associated policies to choose, but also about *how* to choose them. In a world of limited U.S. power, influence, and resources, choices are required. Failure to make decisions is also a choice, and often not a wise one. We offer criteria that may be helpful when making hard decisions in the face of uncertainty: the need to scrutinize assumptions, seize opportunities, uphold U.S. commitments, play both the long and the short games, align U.S. interests with values, and limit regret.

Today's problems are not insurmountable, and opportunities can be exploited. Nor are our times more daunting than those faced by previous generations of Americans. Much of the world still looks to the

---

[24] Ries, Charles P., *Improving Decisionmaking in a Turbulent World*, Santa Monica, Calif.: RAND Corporation, PE-192-RC, 2016.

United States for leadership. The question is whether the country can muster the political will to lead.

In conclusion, we argue that domestic political dysfunction is the greatest obstacle to effective U.S. global leadership. A coherent international strategy will be difficult to purse without a greater degree of domestic political consensus; a "grand strategy" is simply impossible. Building domestic political consensus on the U.S. role in world affairs would require bipartisan agreement to reconcile the competing demands of domestic needs with the requirements of effective international engagement. U.S. leaders would also need to break the political deadlock between funding requirements for the desired degree of international engagement and commensurate levels of taxation. On the basis of such reconciliation, U.S., diplomacy, defenses, and alliance relationships could be adequately resourced to support the chosen strategy. The government could develop coherent plans to promote economic growth, national security, and international engagement in tandem, as required in a world where domestic well-being is inextricably linked to global developments.

Without such political consensus, the United States will face heightened risks.

Part One
Current Challenges

CHAPTER TWO

From Triumph to Stalemate: The Loss of American Consensus

> . . . as we look around the world, we encounter upheaval and conflict. The United States has not faced a more diverse and complex array of crises since the end of the Second World War.
>
> Henry Kissinger[1]

> . . . the march of human progress never travels in a straight line . . . dangerous currents risk pulling us back into a darker, more disordered world.
>
> Barack Obama[2]

The past 13 years, spanning the presidencies of both George W. Bush and Barack Obama, saw a gradual unraveling of a broad American consensus on the goals and purpose of U.S. foreign policy that endured for more than 60 years.

From 1945 to about 2003, despite fierce political battles about specific U.S. policies—most of all, the Vietnam War—the country was largely united in its view of America's role in the world. Fierce political fighting erupted over the appropriate tactics for dealing with pressing

[1] Henry Kissinger, "Global Challenges and the U.S. National Security Strategy," Testimony before the Senate Armed Services Committee, January 29, 2015.

[2] Barack Obama, remarks to the United Nations General Assembly, New York, September 28, 2015.

international problems. All presidents faced opposition inside and outside their political parties in obtaining support for their internationalist policies, none perhaps more than President Harry Truman, who first gave voice to a U.S. postwar foreign policy and had to defend this position against active opposition, including to his decision to go to war in defense of the Republic of Korea. Of course, Vietnam was a wrenching decision that cost Lyndon Johnson his presidency and served as a major source of tension for Richard Nixon and a driver of his efforts to end the confrontation with China. Still, there was broad agreement on the country's three most essential goals: defeating Communism, building a liberal world order that would prevent another world war or nuclear exchange, and increasing global prosperity.

Historians might argue about when this consensus began to unravel, but we see 2003 as an inflection point. Over the course of that year, U.S. allies, other than the United Kingdom, refused to back the United States in its invasion of Iraq; the U.S. military was not able to control the security situation following the collapse of the Saddam Hussein regime; weapons of mass destruction were not found; and al Qaeda waged a campaign of terrorist attacks across the Middle East. Over the next decade, many Americans began to question the wisdom of the invasion, then to harbor increasing doubt about the ability of the U.S. government to "win" Middle East wars or halt terrorism. The situation was made worse in 2008, when the global economy faced the worst financial crisis since the Great Depression. Not only was the quagmire in the Middle East robbing Americans of their sons and daughters along with their treasure, but the global economy, America's greatest hope for the future, appeared on the brink of ruin. During the 2016 presidential campaign, the partisan wounds that had opened over the Middle East wars and deepened during the 2007–2008 financial crisis seemed to fester, the security situation for the United States worsened, inequality and accompanying anger grew, and bipartisan consensus seemed impossible on almost every issue, including foreign policy. The United States was stalemated. The election of Donald J. Trump as the 45th president continued the rancorous era of American politics and for the first time in seven decades threw into question the future of the liberal world order.

This chapter charts the changes in the world during the post–Cold War period. Chapter Three will examine the dramatically altered world we face today.

The End of the Soviet Union and the War That Wasn't

The Cold War ended without a bang. There was no battle, no peace talks, no homecoming parade when the United States became the preeminent world power, militarily, economically, and culturally. The former Soviet Union experienced a humiliating failure as its economy collapsed and its Warsaw Pact satellites lined up to join the North Atlantic Treaty Organization (NATO) and the EU. China, preoccupied with domestic growth and political stability in the aftermath of the Tiananmen Square protests, was largely quiescent on the international stage.

In hindsight, the post–Cold War period may come to be seen as a historical anomaly, an unsustainable moment of dominance by a single power. Nonetheless, several generations of American leaders and voters came of age during this brief time of unchallenged U.S. leadership of an unusually stable world. The performance of subsequent presidents may be judged, however unfairly, by that benchmark. The "millennial generation," on the other hand, came of age during the long and costly wars in Iraq and Afghanistan. Unsurprisingly, millennials tend to see the United States as too involved in world affairs; they favor international cooperation and question the use of force.[3] Both views continue to persist in the body politic.

Prosperity, Peace, and Democracy

The period after World War II brought a global expansion of democracy, prosperity, and freedom—the Pax Americana—and this was

---

[3] In a 2011 Brookings survey of young leaders, born from 1980 to 2005, 57.6 percent said the United States is "too involved in global affairs." Peter W. Singer, Heather Messera, and Brendan Orino, *D.C.'s New Guard: What Does the Next Generation of American Leaders Think?* Brookings Institution, February 2011, p. 14. Also see A. Trevor Thrall and Erik Goepner, *Millennials and U.S. Foreign Policy*, Cato Institute, June 16, 2015.

even more pronounced during the immediate post–Cold War period. Until recently, uninterrupted economic growth may have seemed like the natural state of human affairs, but it is a phenomenon of the past 200 years. In 1820, between 85 percent and 95 percent of the global population are estimated to have lived in poverty.[4] In 1990, according to the World Bank, 37 percent lived in absolute poverty, defined as less than $1.90 per day, and by 2012, this figure had been reduced to 12.7 percent.[5] Global real gross domestic product (GDP) nearly doubled during the post–Cold War period, from $30,924 billion in 1990 to $58,148 billion in 2014, in constant 2005 dollars (Figure 2.1).

Figure 2.1
World Real GDP

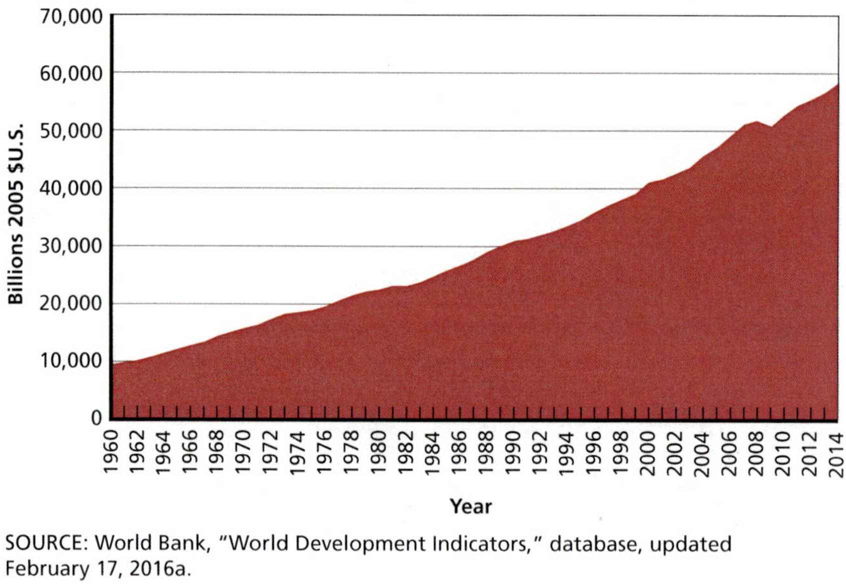

SOURCE: World Bank, "World Development Indicators," database, updated February 17, 2016a.
RAND *RR1631-2.1*

---

[4] Max Roser, "GDP Growth over the Last Centuries," OurWorldInData.org, website, undated-a.

[5] However, the World Bank reports slower progress toward eliminating poverty when a higher standard of poverty, living on $3.10 per day, was applied. World Bank, "Overview," web page, October 7, 2015b.

As anticolonial wars and then Cold War proxy conflicts ended, the incidence of interstate war declined to historic lows in the 1990s (see Figure 2.2). Societal conflicts plummeted as well.[6]

Increasing U.S. and UN diplomatic activism to resolve civil wars, reinforced by UN peacekeeping operations, has contributed to an absolute as well as relative decline in the number of wars.[7]

As shown in Figure 2.3, this period also saw a significant increase in the number of UN stabilization missions.[8]

Under UN Security Council mandate, the United States intervened militarily for largely humanitarian reasons in Somalia, Bosnia, and Haiti. It sought to enforce the principle of territorial integrity, evicting Saddam Hussein's military from Kuwait in 1990. That said,

Figure 2.2
Global Trends in Armed Conflict, 1946–2013

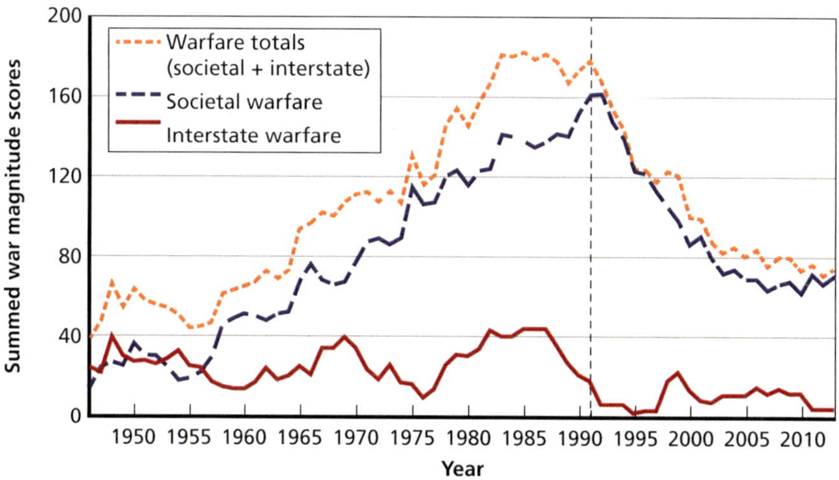

SOURCE: Monty G. Marshall and Benjamin R. Cole, *Global Report 2014: Conflict, Governance, and State Fragility*, Vienna, Va.: Center for Systemic Peace, 2014, p. 13.
RAND *RR1631-2.2*

---

[6] Human Security Centre, *Human Security Report 2005: War and Peace in the 21st Century*, New York: Oxford University Press, 2005, pp. 147–148; Marshall and Cole, 2014.

[7] Human Security Centre, 2005, pp. 147–148.

[8] Human Security Centre, 2005, p. 153.

Figure 2.3
UN Stabilization Operations, 1946–2013

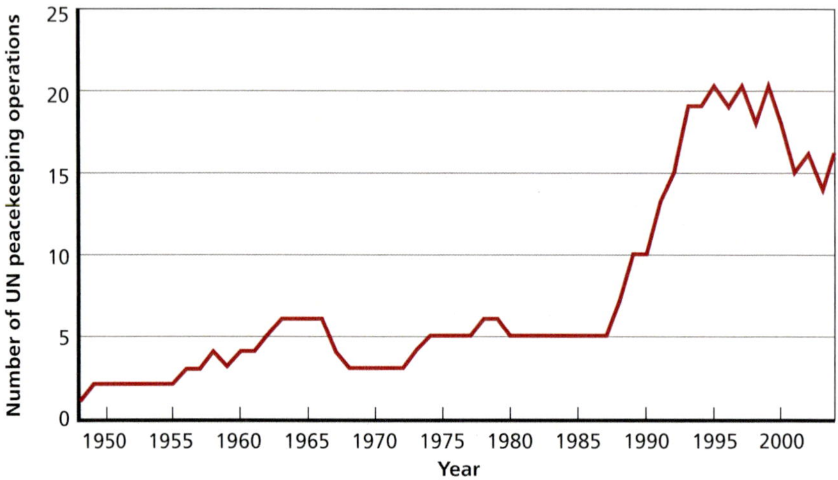

SOURCE: Human Security Centre, 2005.
RAND *RR1631-2.3*

the United States and its allies chose not to intervene in the Rwandan genocide of 1994. Remorse over international failure to act prompted the United Nations to assign a higher priority to the protection of civilians in subsequent missions. More than a decade later, in 2005, the UN General Assembly, meeting at the level of heads of state and government, endorsed the doctrine of R2P, asserting the authority of the international community to intervene when states fail to protect their own populations against mass atrocities.[9] This norm has been slow to take root, however, as its implementation has been limited in many cases by opposition from Russia, China, and even the United States. One example of its successful application is in Côte d'Ivoire.[10]

---

[9] United Nations Security Council Resolution 1894, on protection of civilians in armed conflict, November 11, 2009.

[10] In today's world, numerous social, economic, and security trends are eroding the sovereign powers of national governments, among them the communication capabilities of information technologies (for example, the Internet and mobile phones). Yet such countries as China and Russia seek to promote a norm of "cyber sovereignty"—the ability of state governments to censor international communication flows. The United Nations mounted a peacekeeping mis-

The post–Cold War era was an unusually sunny period for the West. Despite a range of emerging problems—nuclear proliferation in North Korea, Pakistan, and then Iran; the horrors in the Balkans, Rwanda, and then Darfur; the emergence of jihadist movements; and even the messy attempts to contain Saddam Hussein after the first Gulf War—the U.S. homeland and its citizens were largely untroubled by foreign affairs. The institutions that the United States helped build after World War II fostered a liberal international order that promoted free trade, democratic governance, the rule of law, global health, arms control agreements, human rights conventions, and a host of other norms and institutions that promoted peaceful global integration.

The liberal order seemed to be living up to its Wilsonian promise as a time of global peace.[11] The order was opened to all who wanted to participate, and most did: About 3.5 billion people joined the global economy from the previously closed economies of the former Soviet Union, India, and China. Incomes soared, extreme poverty eroded, and the benefits of science, technology, and medicine reached millions who had been left out. A global middle class was born. It was one of the greatest periods of social and economic progress in human history.

Democracy was also on the march. The number of countries deemed "fully free" nearly doubled from 44 in 1973 to 76 in 1992 because of decolonization and the breakup of the Soviet Union, and peaked at 90 in 2007–2008 before falling back to 86 in 2016.[12] The percentage of the global population living in countries that were free or partly free also surged (see Figure 2.4). Beginning in 2006, however, the degree of freedom in many emerging and some estab-

---

sion in Côte d'Ivoire in 2004, but after fighting broke out over a disputed election in 2010, it authorized a mission specifically to fulfill R2P duties. See United Nations Security Council Resolution 1975, on Cote D'Ivoire, March 30, 2011. However, R2P was not invoked under somewhat similar circumstances in Congo. See United Nations Regional Information Centre for Western Europe, "Responsibility to Protect: Two Case Studies with Alex Vines," web page, undated.

[11] Walter Russell Mead, *Special Providence: American Foreign Policy and How It Changed the World*, New York: Taylor and Francis Group, 2009.

[12] Freedom House, *Freedom in the World 2016: Anxious Dictators, Wavering Democracies*, Washington, D.C., 2016a.

lished democracies began to decline, a trend often referred to as a "democracy recession."[13]

Expanding Alliances at Little Short-Term Cost

A notable feature of the immediate post–Cold War years was the expansion of alliances and partnerships. Altogether, the United States has significant security commitments to approximately 140 nations, roughly half of which are highly formalized through treaty obligations.[14]

The United States has excelled in its ability to develop and sustain such relationships as NATO, other major alliances, and multilateral

Figure 2.4
Expansion of Democracy, 1947–2014

The number of world citizens living under different political systems.

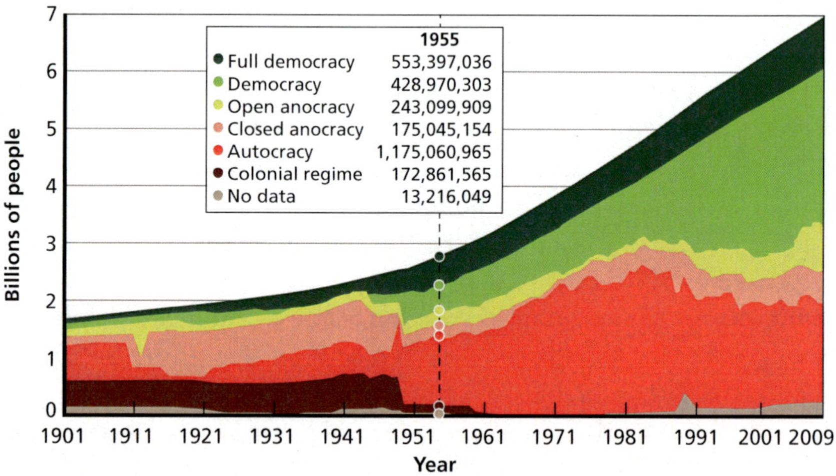

SOURCE: Max Roser, "Political Regime," OurWorldInData.org, website, undated-b.
RAND *RR1631-2.4*

---

[13] According to Freedom House rankings, the number of countries where freedom is declining has outstripped the number where freedom is increasing each year from 2006 to 2016. See Freedom House, 2016a; and Freedom House, "Maps and Graphics," web page, 2016c. See also, Larry Diamond, "Facing Up to the Democracy Recession," *Journal of Democracy*, Vol. 26, No. 1, January 2015.

[14] Jennifer Kavanagh, *U.S. Security-Related Agreements in Force Since 1955: Introducing a New Database*, Santa Monica, Calif.: RAND Corporation, RR-736-AF, 2014.

trade arrangements. Its major competitors, such as Russia and China, are notably lacking in such advantages.

Alliances and partnerships are the outer defenses of the security and economic systems. They support the U.S. economy through greater trade and investment and through the ability to coordinate policy and work through differences in regular forms. By increasing security, they also reduce the costs to U.S. companies of shipping by both air and sea,[15] and they project U.S. political influence in bilateral relationships and through multilateral organizations.[16]

In the aftermath of the Cold War, the United States accumulated security commitments across the globe at a time when it was easy and seemingly cost-free to do so. NATO expanded without a shot fired. Ten countries were admitted, frustrating Russian nationalists while expanding U.S. defense commitments. NATO has brought unprecedented security to its eastern members, although the costs and risks to the alliance from Russian revanchism are yet to be fully assessed. The limits of NATO's enlargement are now being tested.

The Association of Southeast Asian Nations (ASEAN) doubled from five countries at its 1967 founding to ten member states today. It has evolved into a significant force for the growth of trade, as well as economic, political, and security cooperation. Partnerships such as ASEAN are not formal military alliances but imply some degree of U.S. security interest.

An Interconnected World

Eras in international affairs have been defined by the influence of major nation-states—by their acquisition of power, rise to preeminence, competitions with rival states, and wars. The past century was shaped by geopolitical conflicts between the United States and imperial and Communist nations (Germany, Japan, the Soviet Union, and China), and by geo-economic challenges (Japan and China). Our current era is being defined by the emergence of mass publics, groups, and individu-

---

[15] Daniel Egel et al., *Estimating the Value of Overseas Security Commitments*, Santa Monica, Calif.: RAND Corporation, RR-518-AF, 2016, p. 6.

[16] Egel et al., 2016.

als empowered by religious or cultural conviction and information-sharing technologies. This might be called an era of *geo-publics*.

The knitting together of societies and peoples has accelerated to the point where it now triggers backlash among some groups in the United States and other countries.[17] During the Cold War, however, advances in information and communications technologies were seen as inherently redounding to the benefit of the West, which assumed that Communist states could not long survive without propaganda, censorship, and strict control over information from capitalist countries. Shortwave radio during the Cold War gave way to satellite television transmissions during the Nixon administration's opening to China, cassette tapes during the Iranian Revolution, facsimile machines during the Tiananmen Square uprising, television during the final days of the Warsaw Pact and the fall of the Berlin Wall, the World Wide Web, and mobile and Internet telephony. Mass international travel was spurred by visa-free travel for citizens of many countries, more-affordable air travel, the Schengen Agreement that made much of Europe borderless, and the lifting of travel bans in most former Communist and authoritarian nations. The very notion of "globalization" came to mean political, economic, and cultural integration on a historic scale,[18] and promised a new era of international tolerance and respect. Self-isolated regimes such as North Korea, dubbed the "Hermit Kingdom," were not expected to endure.

Economic ties among nations soared with global economic growth and the spread of rules-based trade and investment agreements, such as those through the World Trade Organization (WTO) and the North American Free Trade Agreement (NAFTA). China's accession

---

[17] The 2012 riots in many Muslim-majority countries in response to the "Innocence of Muslims" YouTube video is one example, as is support for Internet censorship of pornography or statements that are offensive to minority or religious groups. The United States is a global outlier in its strong support for virtually unlimited free speech. Richard Wike and Katie Simmons, "Global Support for Principle of Free Expression, but Opposition to Some Forms of Speech," Pew Research Center, web page, November 18, 2015.

[18] Thomas Friedman, *The Lexus and the Olive Tree: Understanding Globalization*, New York: Farrar, Straus and Giroux, 1999.

to the WTO in 2001, capping two decades of economic reform,[19] was a major landmark. Globalization was assumed to be irreversible and of benefit to all nations.

The trade deals that are now so contentious in U.S. politics contributed to a worldwide growth in incomes, if not in wages for some segments of the American workforce. One widely accepted analysis has found that a 1-percent increase in trade relative to GDP leads to growth of per-capita GDP of between 0.85 percent and 1.97 percent.[20] A 2005 study pegged the value of trade to the U.S. economy on the order of $1 trillion, the equivalent of a gain in income of $2,800 to $5,000 each year for the average person in 2003 dollars.[21] U.S. corporations, in particular, reaped the benefits of the new global economy, as did high-skilled workers. The technology boom seemed unstoppable. Even when the "dot-com bubble" burst in 2000, it seemed only temporary, an inevitable correction that would surely be followed by another period of sustained growth. The United States and its friends were filled with confidence.

Twin Shocks: 9/11 and the Financial Crisis

The shift from American triumphalism to anxiety and self-doubt happened on a single day: September 11, 2001. The al Qaeda attacks were a strategic surprise that had effects lasting to this day. After the attacks, most realized the United States would feel compelled to respond to the perpetrators. None anticipated that 15 years later, the United States would be faced with even larger problems: the unraveling of Iraq and

---

[19] It took China 15 years of arduous negotiations to be admitted to the WTO in December 2001. Permanent Mission of China to the WTO, "China in the WTO: Past, Present and Future," briefing, 2011.

[20] Jeffrey A. Frankel and David Romer, "Does Trade Cause Growth?" *American Economic Review*, Vol. 89, No. 3, June 1999.

[21] Scott C. Bradford, Paul L. E. Grieco, and Gary Clyde Hufbauer, "The Payoff to America From Global Integration," in C. Fred Bergsten, ed., *United States and the World Economy: Foreign Economic Policy for the Next Decade*, Washington, D.C.: Institute for International Economics, January 2005.

Syria, the spread of jihadism, and the largest global refugee crisis since World War II.[22]

Just as American voters were concluding that they wanted the United States out of Afghanistan and Iraq, the country was hit with the second shock of the decade: the U.S. financial crisis of 2008 and its spread to economies around the world. It triggered the worst economic crisis since the Great Depression of the 1930s.[23] Eight years later, some national economies are still trying to recover, recovering unevenly, or, in some cases, not recovering. The global contagion compromised confidence in U.S. financial markets and economic management. It demonstrated that regulatory failings and mismanagement in one major country can damage economies around the world. The vulnerabilities associated with globalization became evident again in 2015, as China's growth rate declined, destabilizing the economies of commodities producers around the world and prompting fears of a "hard landing."

The financial crisis had other unwelcome consequences. Russia survived by dint of its currency reserves, and the incident is said to have hardened President Vladimir Putin's determination to chart an economic future independent of the West.[24] Yet Putin failed to make structural reforms that could have increased economic resilience and reduced dependence on volatile commodity prices. Food prices and

---

[22] In this sense, countries can also be surprised by the consequences of their own actions or responses. Did George W. Bush intend to initiate the longest war in American history after September 11, 2001? Did David Cameron intend to trigger Britain's exit from the EU by calling a popular referendum on the question?

[23] In 2008, world GDP grew 1.5 percent, down from 3.9 percent the year before, and then fell by 2.1 percent from 2008 to 2009. The recession was worse in other high-income countries than in the United States, with their collective GDP falling by 3.5 percent between 2008 and 2009, while U.S. GDP fell 2.8 percent.

[24] Hill and Gaddy argue that a strong case can be made that Putin's decision in September 2011 to return to the presidency was motivated by fear of contagion from the financial crisis that was sweeping across Europe (Fiona Hill and Clifford G. Gaddy, *Mr. Putin, Operative in the Kremlin*, Washington, D.C.: Brookings Institution Press, 2015, pp. 85–88 and 246–248). Of course, the financial crisis may have only provided justification for Putin's preexisting opposition to Western economic doctrines. In any case, Putin's economic policies represented an about-face from the integrationist policies of President Dmitry Medvedev.

food insecurity increased around the world. And as a financial crisis that began on Wall Street hurt the livelihoods of lower-income people around the world, income inequality became an increasingly salient political issue in the United States and many other countries.[25]

The theme that prosperous American elites were out of touch with the hardships faced by those left behind by globalization came to dominate the narrative of the 2016 presidential election, culminating in the election of Donald J. Trump as the 45th president of the United States. While the election results were a surprise to many, data documenting the gap between the two Americas had been accumulating for some time. In 2012, for example, Charles Murray laid out the vast and growing economic and cultural divide between educated whites living in affluent communities and their working-class counterparts living in more-distressed neighborhoods that are increasingly segregated by social class.[26] Four years later, an analysis of distressed and affluent zip codes by the Economic Innovation Group showed that despite the nominal recovery from the Great Recession, rich and poor communities had pulled even further apart, with 50.4 million Americans living in communities plagued by poverty, lack of education, and joblessness.[27] The wealthiest, most-educated segments of American society recovered and even prospered after the Great Recession, while the least-advantaged parts of American society continued to lose ground, caught in an almost decade-long stall with few or no prospects for progress in sight.

---

[25] In October 2015, the Organisation for Economic Co-operation and Development (OECD) began a major project to study increasing inequality and possible solutions. See OECD, "Income Inequality and Poverty," web page, undated-b.

[26] Charles Murray, *Coming Apart: The State of White America 1960–2010*, New York: Crown Forum, 2012.

[27] In the one-fifth of distressed zip codes, 23 percent of adults have no high school degree, the poverty rate is 27 percent, 55 percent of adults are not working, and income is 69 percent of the national median. In the top quintile of zip codes, by contrast, only 6 percent lacked a high school degree, the poverty rate was 6 percent, 35 percent of adults were not working, and the median income was 146 percent of the national sum. Economic Innovation Group, "The 2016 Distressed Communities Index: An Analysis of Community Well-Being Across the United States," February 25, 2016.

Sources of American Self-Doubt

American ambivalence and division about the U.S. role in the world seems to stem in part from the high costs and inconclusive results of the wars in Afghanistan, Iraq, and against Salafi-jihadism.[28]

As suggested by Figure 2.5, satisfaction with the U.S. position in the world appears to have little to do with the severity of national security threats and more to do with public perceptions of the effectiveness of the government response to such threats. This trend is evident in earlier surveys, as well. For example, in 1963, only 44 percent of Americans were dissatisfied with the country's position the world,[29] although the Cuban Missile Crisis had demonstrated the considerable existential danger posed by Soviet nuclear weapons. And 71 percent of those surveyed were satisfied with America's role in the world in 2002, despite warnings that al Qaeda could be preparing more attacks on the homeland.

American self-doubt—doubt about U.S. international effectiveness combined with doubt about whether many citizens can still achieve the "American Dream"—has itself become a constraint on engagement abroad. This self-doubt has many sources. One stems from the government's inability to achieve its major foreign policy goals, including defeating radical jihadism, ending the war in Syria, and ensuring that Iran cannot acquire nuclear weapons. A second stems from fears that the U.S. government will be unable to distinguish terrorists from migrants and an associated fear of immigrants from Syria and other Muslim nations, which became a prominent issue in the 2016 election campaign.[30]

A third source of American self-doubt is financial: How can the country deal with global terrorism, an anemic economy, and mounting debt all at the same time? Federal debt in relation to the size of the

---

[28] For example, in an April 2016 survey, 62 percent of voters approved of the U.S. military campaign against ISIS but 58 percent felt it was not going well. The percentage of those who thought the United States was doing "fairly well" in reducing the threat of terrorism fell from 88 percent in 2001 to 54 percent. Pew Research Center, 2016b, pp. 32, 36.

[29] The Gallup Organization has asked this question on the "U.S. Position in the World" at regular intervals from July 1962 to February 2016.

[30] Smeltz et al., 2016.

Figure 2.5
Public Dissatisfaction with America's Role in the World, 2000–2012

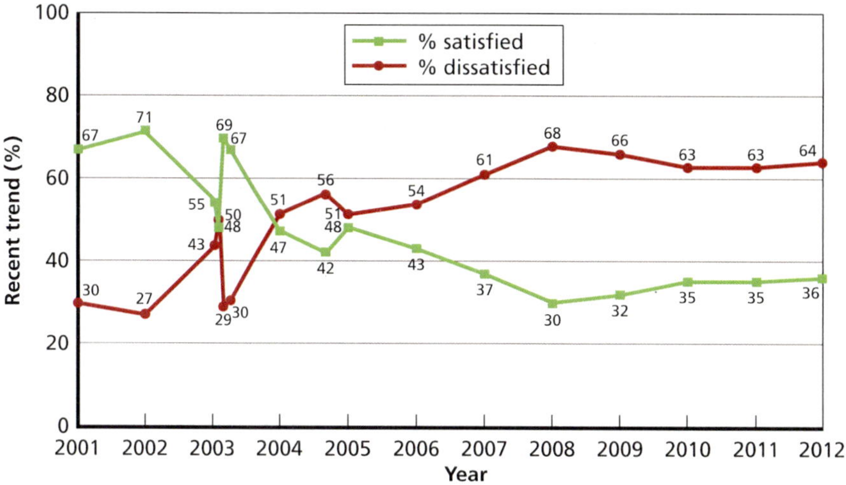

SOURCE: "U.S. Position in the World," 2016.
RAND *RR1631-2.5*

economy has doubled since 2000. As of 2015, debt was 74 percent of GDP. As shown in Figure 2.6, the Congressional Budget Office (CBO) estimates that it could exceed GDP in 2040.

The highest debt level in history was in 1946, when debt peaked at 106 percent of GDP because of the money borrowed to fight World War II.[31] The Iraq and Afghanistan wars cost at least $1.8 trillion in direct military spending from 2001 to 2015.[32] The United States did not raise taxes to fund these conflicts and will continue to pay the interest costs of borrowing the funds for those wars. Political gridlock in Congress led to a down-to-the-wire battle over whether the country,

---

[31] CBO, 2016, pp. 14–15; Shatz, 2016, pp. 23–25.

[32] This figure understates the total war costs because it represents the cumulative total of the Overseas Contingency Operations (OCO) budgets from 2001 to 2015, in 2016 dollars, which totaled almost $1.8 trillion, but does not include medical care or other veterans' benefits or other indirect costs. OCO spending averaged $117 billion per year during this period. For fiscal year 2016, the U.S. Department of Defense (DoD) requested $51 billion for OCO. CBO, 2016, pp. 14–15.

Figure 2.6
Debt as a Percentage of GDP

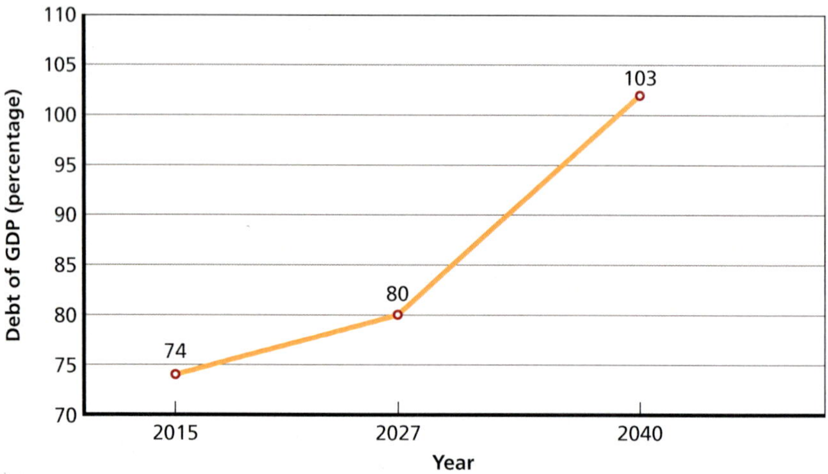

SOURCE: Howard J. Shatz, *U.S. International Economic Strategy in a Turbulent World*, Santa Monica, Calif.: RAND Corporation, RR-1521-RC, 2016, Figure 2.10, updated with data from CBO, *Long-Term Implications of the 2016 Future Years Defense Program*, Washington, D.C., CBO 51050, January 14, 2016, pp. 14–15.
NOTE: When will servicing debt crimp U.S. policy options? CBO estimates that in 2040, interest payments could absorb more than 20 percent of all federal revenue, gobbling up funds that could be invested in domestic advances and international efforts.
RAND *RR1631-2.6*

for the first time in its history, would default on its debt. In 2011, Standard and Poor's downgraded the U.S. credit rating from AAA to AA+, where it remains, despite an improving economy, because of high U.S. debt levels. If the United States remains on its current course, CBO estimates that entitlement spending could amount to 73.2 percent of GDP by 2040, while interest on the debt could be 4.3 percent—an unsustainable path.[33]

A fourth source of American self-doubt involves the conduct of modern warfare and its effects on local populations and humanitarian norms. While both wars brought quick, promising results—recall the

[33] Keith Hall, *The 2015 Long-Term Budget Outlook*, Washington, D.C.: Congressional Budget Office, June 16, 2015.

images of U.S. Special Forces riding on horseback to apparent victory in Afghanistan, or the falling statue of Saddam Hussein in Baghdad—the results were fleeting. Insurgent activities grew, casualties mounted, and atrocities came to light. International criticism over the conduct of the wars dented U.S. credibility on matters of human rights and international law. The searing images from the Abu Ghraib prison in 2004 triggered domestic and international criticism; a decade later, similarly searing images of Iraqi and Syrian casualties and desperate refugees triggered international horror but little remedial action. Despite an early sense of achievement that came with both wars, the lingering effects have been of disappointment, frustration, and even failure. Doubts soared regarding the ability of U.S. forces to succeed in far away conflicts.

The opportunity costs of the Afghanistan and Iraq wars are not easy to measure. They remain the subject of acrimonious debate in which the cost of the Iraq War has been compared with the opportunity to fund preschools, college tuitions,[34] or more investment in physical infrastructure.[35]

Meanwhile, there is growing national debate over a wide range of indicators that a segment of American society is failing to thrive, partly because of socioeconomic problems that have not been mitigated by recovery from the Great Recession. The "downward mobility" of some formerly middle-class Americans is more than an economic phenomenon. Americans also now have shorter life expectancy and worse health outcomes than their counterparts in other affluent nations.[36] Mortality

---

[34] Student loan debt totals more than $1.2 trillion, owed by 41 million Americans, with more than one in four borrowers in delinquency or default. Consumer Financial Protection Bureau, *Student Loan Servicing: Analysis of Public Input and Recommendations for Reform*, Washington, D.C., September 2015.

[35] American Society of Civil Engineers, "Report Card for America's Infrastructure," web page, 2013. While civil engineers, it should be acknowledged, are an interested party, the report estimated that $3.6 trillion in infrastructure investments were required by 2020, slightly higher than the then-estimates of the cost of the wars in Iraq and Afghanistan.

[36] Mauricio Avendano and Ichiro Kawachi, "Why Do Americans Have Shorter Life Expectancy and Worse Health Than People in Other High-Income Countries?" *Annual Review of Public Health*, Vol. 35, 2014; Elizabeth Arias, "Changes in Life Expectancy by Race and Hispanic Origin in the United States, 2013–2014," Centers for Disease Control and Prevention, NCHS Data Brief, No. 244, April 2016.

rates of white males ages 45 to 54 increased by half of 1 percent after 1998, attributed in part to suicide and opioid and alcohol abuse.[37] The least-educated white Americans were particularly affected.[38] Notably, the U.S. suicide rate has increased this century, jumping by more than 28 percent between 1999 and 2010; in 2009, the peak of the recession, suicides outnumbered deaths from car accidents.[39]

Abrupt changes in mortality rates are typically seen in countries during times of war or economic crisis, such as in Russia following the collapse of the Soviet Union and in Africa during the HIV/AIDS crisis prior to the introduction of antiretroviral therapies. Taken *in toto*, these negative trends in the well-being of a substantial segment of the population will remain a constraining factor in U.S. foreign policy, perhaps for the next several presidential administrations. American voters are asking how U.S. foreign policy will advance their interest, not just global interests.

Today, the "post–Cold War period" is over. Historians may argue whether it ended with the al Qaeda attacks of September 11, 2001, or perhaps on November 5, 2008, when Obama was elected President after declaring that the U.S. response to 9/11 had been a failure. Whatever the timing, it has become clear to most foreign policy practitioners that the world has entered a new era, a complex age of turbulence and opportunity. The challenge of this century is not just to hold the gains of the last century, but to build upon them.

---

[37] Anne Case and Angus Deaton, "Rising Morbidity and Mortality in Midlife Among White Non-Hispanic Americans in the 21st Century," *Proceedings of the National Academy of Sciences*, Vol. 112, No. 49, December 8, 2015.

[38] See, for example, Betsy McKay, "The Death Rate Is Rising for Middle-Aged Whites," *Wall Street Journal*, November 3, 2015, and Sabrina Tavernise, "Life Spans Shrink for Least Educated Whites in the U.S.," *New York Times*, September 20, 2012.

[39] Erin M. Sullivan et al., "Suicide Among Adults Aged 35–64 Years—United States, 1999–2010," Centers for Disease Control and Prevention, *Morbidity and Mortality Weekly Report*, Vol. 62, No. 17, May 3, 2013, pp. 321–325.

CHAPTER THREE

Is the World Falling Apart (And How Would We Know)?

> The nature of power is changing. The nature of international cooperation is changing. The nature of conflict is changing. We're not evolving well to adapt. This world is not as dangerous as that during the Cold War, but it is *much* more complicated.
>
> Former national security adviser Brent Scowcroft[1]

Strategists succeed when their assumptions prove correct, and when the world they plan for is the world that they encounter. What kind of world is emerging in the 21st century?

We anticipate an environment of turbulence, complexity, and contradiction. The world is being bound together by instant communications, commerce, and shared knowledge, yet fragmenting around major state rivalries, contrasting visions of world order, conflicting values, sectarian feuds, and aggressive nationalist passions. Geopolitical realignments will challenge the writ of strong states and the survival of fragile ones. Transnational trends are challenging traditional notions of state sovereignty. Political strife over income inequality, the global competition for jobs, and mass migration will continue to challenge the logic of globalization. Urbanization will intensify, and water shortages are expected.[2]

---

[1] Quoted in Rothkopf, 2014, p. 350.

[2] Office of the Director of National Intelligence, *Global Trends 2030: Alternative Worlds*, Washington, D.C.: National Intelligence Council, 2012.

Technology, as always, will be a key driver of change. The socio-political effects of technological changes in earlier eras—for example, following the introduction of electricity, radio, telephones, automobiles, airplanes, and the atomic bomb—were just as sweeping.[3] Some point to the "creative destruction" that is reshaping the global economy and destroying a way of life for many Americans who are losing jobs in the "old" economy and floundering in the new one. But the rapid transition from agricultural to industrial societies in the last century was no less disruptive. The question is not whether the displacement of people by rapid societal and technological change is more acute than in times past, or whether the rate of global change is accelerating or not. It is simply too soon to tell. In the meantime, the policy challenge is how to reeducate and reintegrate dislocated workers and increase the supply of higher-wage jobs as a means to increase citizen well-being—and with it, domestic political tranquility and national cohesion.

In the developing world, growth is exploding, yet political and social fragility is ever more pronounced. As *New York Times* columnist Thomas L. Friedman put it:[4]

> Many of these fragile, artificial states don't correspond to any ethnic, cultural, linguistic or demographic realities. They are caravan homes in a trailer park—built on slabs of concrete without real foundations or basements—and what you're seeing today with the acceleration of technology, climate change stresses and globalization is the equivalent of a tornado going through a trailer park.

Powerful new technologies will be developed and put to both benign and malignant uses, as they always have been. These advanced technologies will be available to more individuals and more states. The United States will need to work harder to address this conundrum, by attempting to restrict the proliferation of potentially dangerous technologies to unfriendly states and nonstate actors, while at the same

[3] See Neil Irwin, "What Was the Greatest Era for Innovation? A Brief Guided Tour," *New York Times*, May 13, 2016.

[4] Thomas L. Friedman, "What If?" *New York Times*, January 20, 2016.

time promoting the dissemination of appropriate ones—along with education and public policy guidance—to advance development and improve global health. But this will require staying one step ahead of the many "tornados" hitting the developing world.

Technological change is also transforming how we think about and prepare for warfare. As Carl Philipp Gottfried von Clausewitz observed, "war's nature doesn't change, only its character."[5] The nature of war has always been and remains violent and political. But the character of war—its conduct—is profoundly and visibly influenced by technology.

Powerful capabilities that were once the exclusive preserve of states are increasingly available to individuals, raising new questions about what kind of nonproliferation regimes are feasible. In an era when anyone can build a drone from a hobby kit and fly it with a smart phone, it is no longer clear what "counterproliferation" or "arms control" will mean. These issues are unlikely to get easier as the human-plus-machine model of warfare develops.

Future administrations will almost certainly find themselves struggling to keep up with (let alone regulate) new information technologies, bio- and nanotechnology, powerful gene-editing tools such as clustered regularly interspaced short palindromic repeats (CRISPR), advances in hacking and encryption, drones, robotics, 3-D printing of weapons, artificial intelligence, and other game-changing developments that could reshape the international economy, politics, and the military balance of power, as they have done throughout history. Preserving overwhelming U.S. military superiority—should the country choose to do so—will be neither cheap nor easy.

---

[5] Clausewitz quoted in Christopher Mewett, "Understanding War's Enduring Nature Alongside Its Changing Character," *War on the Rocks*, blog, January 21, 2014. Lynn E. Davis, Michael J. McNerney, and Daniel Byman, "Armed Drone Myth 1: They Will Transform How War is Waged Globally," The RAND Blog, February 17, 2015.

A Rubik's Cube of Realignments

Whether or not the world is falling apart, the United States is, in many ways, in an enviable position today compared with its rivals. First, the nation faces no certain existential threat. It does face *potential* existential threats from nuclear-armed Russia and China, but without the "we will bury you"[6] animus that put the fear of hot war into the Cold War confrontation with the Soviet Union. Today, these rival states are both adversaries and partners—though recently both appear to be moving more firmly into the adversary camp. Still, the only unalloyed U.S. adversaries are North Korea and violent jihadist movements, as expressed by ISIS, al Qaeda, and related groups.

Nevertheless, this new era presents consequential challenges to U.S. interests and will require the United States to adapt to significant international shifts and realignments.

Major-power realignments include the following:

- **Russia** had become a closer strategic partner to the United States under President Medvedev, with the New START Treaty in 2011 and cooperation in allowing Russian railways to forward NATO supplies to its forces in Afghanistan. As recently as 2010, Washington tried in vain to "reset" relations with Moscow. Yet counterterrorism cooperation with Russia has been uneven, if not unhelpful. The Kremlin's use of military force in Ukraine in 2014 triggered the imposition of U.S. and EU sanctions. Its military intervention in the Syrian conflict in 2015 and 2016 has led U.S. officials to conclude that Russia intends to challenge the United States on a wider scale,[7] with China, Iran, and perhaps Turkey as its allies. The U.S. intelligence community's assessment that Russia attempted to influence the U.S. presidential election

[6] This remark by Soviet Premier Nikita Khrushchev in 1956, was translated as, "Whether you like it or not, history is on our side. We will bury you." Many scholars argue that what Khrushchev meant was "We will outlast you." See Philip H. Gordon, *Winning the Right War: The Path to Security for America and the World*, New York: Henry Holt, 2007, p. 174.

[7] Lt. Gen. Vincent Stewart, director of the Defense Intelligence Agency, testimony to the U.S. House of Representatives Foreign Affairs Committee, March 2, 2016.

in favor of a Trump victory, and subsequent imposition of U.S. sanctions,[8] creates unprecedented challenges.

- **China**, which was no longer considered to pose a military threat to the United States after President Nixon initiated normalization of relations in 1972, has now become both a major economic partner and a regional security challenger. It shares some security interests, particularly in constraining North Korea's nuclear program and maintaining access to global energy supplies. A U.S.-China partnership is seen as indispensable to combating climate change, and the two countries signed a major climate accord in 2014. At the same time, China opposes U.S. plans to deploy antimissile batteries to South Korea and efforts to develop deeper security ties in East Asia. Its territorial aggressions in the South China Sea challenge norms of international law and conflict management and the security interests of longtime allies and partners. China's refusal to enforce new international sanctions against North Korea has allowed North Korea's nuclear weapons capabilities to advance.
- **Europe** can no longer be judged to be stable and at peace. The continent faces deep economic stagnation, a refugee and migration crisis, and internal frictions that raise questions about the future of the EU. Following the United Kingdom's June 2016 "Brexit" referendum in favor of withdrawing from the EU—a step the Obama administration opposed—the country remains internally divided and estranged from its traditional allies. Right-wing parties in many other EU countries also want to quit the union.[9] Some consider the biggest beneficiary of European disunion to be Putin.[10] European values are under attack, and anti-Semitism is on the rise in a number of European countries. Parts of Eastern Europe are moving sharply toward authoritarian gov-

---

[8] White House, "Statement by the President on Actions in Response to Russian Malicious Cyber Activity and Harassment," Washington, D.C., December 29, 2016b.

[9] Kate Lyons and Gordon Darroch, "Frexit, Nexit or Oexit? Who Will be Next to Leave the EU?" *The Guardian*, June 27, 2016.

[10] Michael McFaul, "How Brexit Is a Win for Putin," *Washington Post*, June 27, 2016.

ernment. Russia's threat to Europe's eastern periphery only adds pressure to the already vulnerable region.

- **Turkey**, a NATO ally, has become more authoritarian in the wake of the 2016 coup attempt. With U.S.-Turkish ties fraying, President Tayyip Erdogan has moved to improve ties with President Putin.

Realignments in Asia include:

- The ever-belligerent **North Korea** continues to test its nuclear weapons and the intercontinental ballistic missiles (ICBMs) to deliver them. Beijing has reaffirmed its basic support for the North Korean regime, despite stand-offish relations with the young leader Kim Jong-un.
- **South Korea**'s former enemy, China, has become its top trading partner, but relations have been fraying as Seoul weighs deployment of the terminal high-altitude area defense anti-ballistic missile system from the United States to defend against North Korea.
- **The Philippines,** under President Rodrigo Duterte, has improved relations with China and Russia and sharply criticized the United States. While the United States is very popular in the country, Duterte's stance has cast some doubt on the ultimate direction of the treaty alliance.
- **Vietnam**, threatened by China's maritime claims, is strengthening security ties with Japan and the United States. Other ASEAN countries and Australia are also working with the United States to improve their maritime capabilities.
- **Japan** is increasing military spending and forging stronger defense relationships with Vietnam, Burma, the Philippines, and Australia, to balance China's influence. Japan's recent apology to South Korea for the use of "comfort women" opens the door to more security cooperation between the two countries. Concerned about China's maritime activities, Japan has been pushing a "Diamond Concept" in which the United States, Japan, Australia, and India would form the four points of a diamond-shaped security zone in the Indo-Pacific.

- **India** has long pursued a nonaligned foreign policy, but New Delhi and Washington share concerns over the Chinese naval build-up in the Indo-Pacific. At the same time, India is expanding security cooperation with Japan, cooperating on defense technology, intelligence-sharing, and upgrading infrastructure in the Andaman Islands, while retaining close ties to Russia, India's traditional weapons supplier.

Realignments in the Middle East include the following:

- **Iran** has been improving ties with Russia and China, but its relationship with the United States remains tense and uncertain given Tehran's destabilizing actions in Syria, Iraq, and elsewhere within the region. The Obama administration bought time with the July 2015 agreement to constrain Tehran's nuclear program and lifting of UN and U.S. economic sanctions. The issue, however, is at best deferred and hardly resolved. Future presidents will almost certainly be contending with Iran's nuclear and political ambitions and all that they portend for the highly volatile region. Meanwhile, Tehran could use its enhanced regional standing, newly unfrozen assets, and relationships with Hamas and Hezbollah to destabilize regional governments, or it could play a helpful role in defeating ISIS—or both.[11]
- **Saudi Arabia**, which represents important elements of the Sunni world, sees the United States as an uncertain security partner at best and in collusion with Iran at worst. Yet the Obama administration has largely backed Saudi Arabia's intervention in Yemen. President-elect Trump has suggested taking a harder line on Saudi Arabia

[11] The United States could continue to use economic sanctions that were not covered in the P5+1 deal to punish Iran for support of terrorism and the development of nuclear missiles. (P5+1 refers to the UN Security Council's five permanent members (P5)—China, France, Russia, the United Kingdom, and the United States—plus Germany). Radha Iyengar and Rebecca Friedman Lissner, "Iran, Terrorism, and Nonproliferation After the Nuclear Deal," *War on the Rocks*, January 28, 2016.

- **Israel**'s relationship with the United States is highly strained following the nuclear agreement with Iran and continuing differences over promotion of a two-state solution to the conflict with the Palestinians.

Dealing with these shifting relationships will require deft American diplomacy: the ability to manage ambiguous relationships with governments that have both shared and conflicting interests. Diplomats have done so before, in dealing with the Soviet Union, Saudi Arabia, Pakistan, and many other countries. Creative, multidimensional thinking—"Rubik's Cube" diplomacy—will be all the more necessary in the future. Managing the domestic political reactions to these difficult relationships will continue to be a challenge.

As outlined in previous volumes of the Strategic Rethink series, the United States faces serious challenges from Russia and China.[12] Each has made territorial assertions that challenge international conventions of state sovereignty and legal approaches to managing disputes. And each has ramped up defense spending and made notable military advances that challenge U.S. deterrence.

Russia has emerged as a revanchist power since Putin's return to the presidency in 2012. It now threatens to destabilize Europe through pressure and subversion. As one observer put it, Putin advances Russian power by "creating chasms when possible and living in them."[13] Russia has used force against Ukraine and Georgia, annexed Crimea, taken *de facto* control of an area in eastern Ukraine, and sought to intimidate the Baltic states—Estonia, Latvia, and Lithuania—as well as Poland. To the south, it has expanded its Black Sea fleet. To the North, it has laid claim before the United Nations to more than 463,000 square miles of continental shelf in the Arctic,[14] and created

---

[12] David Ochmanek et al., *America's Security Deficit: Addressing the Imbalance Between Strategy and Resources in a Turbulent World*, Santa Monica, Calif.: RAND Corporation, RR-1223-RC, 2015.

[13] DoD official, discussion with RAND researchers, October 2016.

[14] Andrew E. Kramer, "Russia Stakes New Claim to Expanse in the Arctic," *New York Times*, August 4, 2015.

a new military command for the region. To the West, Russia now confronts NATO with a possible *coup de main* in the Baltics.[15] Moscow's escalation of the war in Syria in support of President Bashar al-Assad, particularly Russian bombing of anti-Assad forces supported by the United States, is the first proxy war between the two countries since the Cold War, with the potential to provoke a dangerous military clash between nuclear powers.

Russia gains a degree of immunity for its aggression from its substantial nuclear arsenal, which it is upgrading. Its "hybrid warfare" tactics, a combination of political and covert subversion, along with deniable military and cyber-operations and information warfare, are more difficult to deal with because of these nuclear capabilities. In the long term, Putin's Russia may well be in decline due to demographic weakness, overdependence on oil exports, a corrupt and inefficient oligarchical system, and a brain drain. However, declining powers can sometimes be the most dangerous.

China has been modernizing its military for two decades. Notably, the growth in its defense spending has almost always outpaced even the spectacular growth of its economy.[16] Much of these ever-growing defense budgets have been used to modernize China's conventional forces and military organization. China has improved its air and missile capabilities (see Figure 3.1), modernized and expanded its navy,

---

[15] RAND has studied the probable outcome of a Russian invasion of the Baltic states, including through repeated wargames conducted in 2014 and 2015, and found that as currently postured, NATO could not successfully defend the territory of its most-exposed members. Russian forces could reach the outskirts of the Estonian capital Tallinn, and/or the Latvian capital, Riga, in less than 60 hours. Researchers concluded that "such a rapid defeat would leave NATO with a limited number of options, all bad." David A. Shlapak and Michael Johnson, *Reinforcing Deterrence on NATO's Eastern Flank: Wargaming the Defense of the Baltics*, Santa Monica, Calif.: RAND Corporation, RR-1253-A, 2016, p. 1.

[16] Roger Cliff writes, "China's defense expenditures as a whole . . . grew at a 12.40 percent rate [between 1999 and 2009]. . . . [This] was higher than the growth rate of China's economy as a whole, which grew at a 10.3 percent annual rate during this period." Roger Cliff, *China's Military Power: Assessing Current and Future Capabilities*, Cambridge, UK: Cambridge University Press, September 2015, pp. 88–89. See also Richard A. Bitzinger, "China's Double-Digit Defense Growth: What It Means for a Peaceful Rise," *Foreign Affairs*, March 19, 2015.

Figure 3.1
China's Missile Capabilities

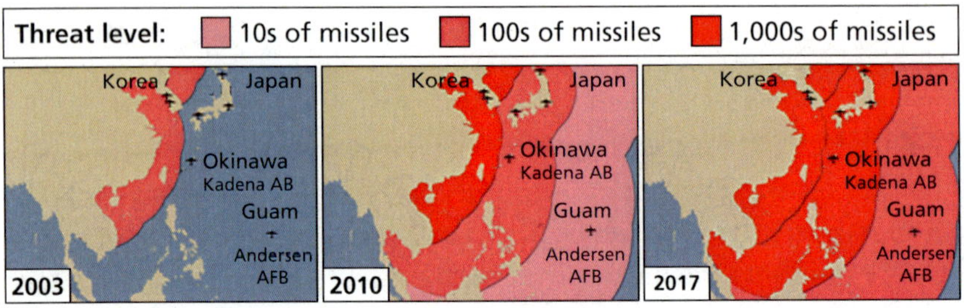

SOURCE: Eric Heginbotham et al., *The U.S.-China Military Scorecard: Forces, Geography, and the Evolving Balance of Power, 1996–2017*, Santa Monica, Calif.: RAND Corporation, RR-392-AF, 2015, p. 51.
NOTE: AB = airbase; AFB = Air Force Base.
RAND *RR1631-3.1*

and adopted more-aggressive postures, particularly in the South China Sea. The United States and China disagree on the right of a nation to regulate foreign military operations within its 200-mile exclusive economic zone, with Washington insisting on the right to free navigation in those waters. China is improving its anti-access/area denial (A2/AD) capabilities to counter the U.S. ability to project military power throughout East Asia, making it less certain that the United States would win a decisive victory in case of war.[17] As part of the buildup, China is constructing thousands of longer-range missiles that could strike targets throughout the region, including major cities such as Tokyo, and even U.S. forces at Andersen Air Force Base in Guam.[18]

Finally, over the last dozen years, China has invested in a modern nuclear weapons arsenal that provides it with an increasingly secure strategic deterrent. It has hardened its ICBM missile sites, dispersed its weapons, and developed the capability to move its ICBMs around the country aboard trucks. Some of these ICBMs carry multiple warheads, which make them very difficult to intercept with missile defenses.

---

[17] David C. Gompert, Astrid Cevallos, and Cristina L. Garafola, *War with China: Thinking Through the Unthinkable*, Santa Monica, Calif.: RAND Corporation, RR-1140-A, 2016.

[18] Heginbotham et al., 2015.

Moreover, in the past year, China has put some of its nuclear forces to sea aboard a new generation of submarines (Type 094 JIN-class) that carry a wholly new class of ballistic missiles (JL-2s).

This combination of hardening, dispersal, mobility, and penetrability significantly reduces the probability that even the most massive and accurate of strikes (conventional or nuclear) would destroy the entirety of China's nuclear force. In short, the modernization of China's nuclear weapons arsenal would appear to provide Beijing with a key capability that it has long sought and was lacking as recently as the mid-1990s: a survivable and thus assured retaliatory nuclear capability.[19] While disconcerting, this capability is not automatically bad for the United States because to the extent that it reassures Beijing, it reduces any incentive for a Chinese first strike.

In terms of its standing in the global economy, China's economic power will not outstrip Washington's in the short term, certainly not if its current downturn persists. Its three decades of state-managed juggernaut growth have created overinvestment in industry and public works, meager domestic consumption, capital misallocation, and an overhang of debt. The stability of its banking sector is in question because Beijing requires state banks to lend to insolvent state corporations and regional and local governments, leading to mounting nonperforming loans. Imprudent interference in stock market performance has reduced confidence in the quality of government economic management.

As its growth rate declines, trading partners from Brazil to Australia whose economies depend on exporting raw materials to China see their development strategies called into question. Yet China remains a major investor in and lender to the developing world, and its "One Belt

---

[19] For in-depth treatments of China's quest for and realization of a survivable modern nuclear deterrent force, see Fiona S. Cunningham and M. Taylor Fravel, "Assuring Assured Retaliation: China's Nuclear Posture and U.S.-China Strategic Stability," *International Security*, Fall 2015; Michael S. Chase, "China's Transition to a More Credible Nuclear Deterrent: Implications and Challenges for the United States," *Asia Policy*, July 2013; M. Taylor Fravel and Evan S. Medeiros, "China's Search for Assured Retaliation: The Evolution of Chinese Nuclear Strategy and Force Structure," *International Security*, Fall 2010; and Thomas J. Christensen, "The Meaning of the Nuclear Evolution: China's Strategic Modernization and U.S.-China Security Relations," *Journal of Strategic Studies*, August 2012.

One Road" program and mercantilist foreign policy are likely to create tensions with the United States and other countries in the years to come.

Even a China that is growing at 5 percent or 6 percent per year will be a major force in the global economy for decades to come. The United States has a strong national interest in ensuring that China does not become a foe, or draw closer to Russia against the United States.

Unlike in the United States, defense spending in China and Russia is not subject to democratic review. Both countries could sustain their military buildups for some time despite economic slowdowns, although fiscal constraints could eventually be a limiting factor.

Empowered Authoritarians

As the post–Cold War period drew to a close, the world witnessed a new confidence among authoritarian regimes[20]—including Russia, China, Turkey, Hungary, Ecuador, and Venezuela—as well as an upswell of populist nationalism in many countries, including the United Kingdom and the United States.[21]

The assertive defense of illiberal values by Moscow and Beijing coincides with their efforts to challenge U.S. alliance relationships and military dominance. Both governments are actively opposing universal human rights norms as well as American values. For example, the Chinese Communist Party and government are countering U.S. appeals for democratic governance and policies supportive of human and civil rights with domestic and international promotion of the Confucian tradition of "harmonious" authoritarian governance. Party cadre, educational institutions, and policy think tanks are instructed to observe "the seven no's:" no discussion of universal values, freedom of speech, crony capitalism, judicial independence, civil rights, civil society, or

---

[20] Christopher Walker, "The Authoritarian Threat: The Hijacking of Soft Power," *Journal of Democracy*, Vol. 27, No. 1, January 2016.

[21] Ronald F. Inglehart and Pippa Norris, *Trump, Brexit, and the Rise of Populism: Economic Have-Nots and Cultural Backlash*, Cambridge, Mass.: Harvard University Kennedy School of Government Faculty Working Paper, RWP16-026, August 2016.

historical errors of the Communist Party.[22] University textbooks are forbidden to use teaching materials that disseminate Western values.[23] And the government has initiated a worldwide effort to promote China's traditional values through Confucian Institutes, media outreach, and online educational programs.

In Russia, Putin appeals for domestic support on the basis of national history and traditional culture, in opposition to the asserted corrupting effects of "Western" values. In 2014, he promoted a "Year of Culture" that sought to assert Russia's unique identity, its cultural roots, values, and ethics.[24] As one recent demonstration of the polarizing effects of a clash of values, President Obama absented himself from the Sochi Olympic Games in 2014 because of Russia's adoption of laws that the United States viewed as infringing on freedom of expression and the rights of lesbian, gay, bisexual, and transgender (LGBT) people.[25] Russia, as well as China and other authoritarian states, have expelled foreign nongovernmental organizations that advocate human rights, electoral politics, and the strengthening of civil society.[26]

In the Middle East, the current political turmoil is driven by fundamental conflicts over religious identity and social values, and popular support for Sharia law challenges the secular underpinnings of international human rights law. Authoritarian governments and hard-

---

[22] Chinese Communist Party Central Committee, "Communiqué on the Current State of the Ideological Sphere," Asia Society China File translation, November 8, 2013; Benjamin Carlson, "7 Things You Can't Talk About in China," *Global Post*, June 3, 2013.

[23] Chris Buckley, "China Assails Textbooks Sowing 'Western Values.'" *New York Times*, January 31, 2015.

[24] Mark Galeotti and Andrew S. Bowen, "Putin's Empire of the Mind," *Foreign Policy*, May/June 2014.

[25] The 2013 law banned "promotion of non-traditional sexual relations among minors." The State Department said it "effectively criminalizes public expression and assembly for anyone who would advocate LGBT equality" while other laws against extremism were used to prosecute religious minorities and make "offending the religious feelings of believers" a criminal offense. U.S. Department of State, Bureau of Democracy, Human Rights and Labor, *Russia 2013 Human Rights Report*, Washington, D.C., 2013.

[26] Asia Society China File, "The Future of NGOs in China," May 14, 2015.

line Islamist groups alike have rejected U.S. efforts to promote democratic governance, respect for girls' and women's rights, and other practices that confront local traditions of suppression of political dissent, sectarian intolerance, male social dominance, female subordination, and home sequestration. The United States is finding its relationships with two longtime and important allies, Turkey and Saudi Arabia, increasingly uncomfortable following Ankara's harsh response to the 2016 coup and the Saudi promotion abroad of its Wahabbist religious doctrine. Congress voted to override President Obama's veto of a bill that would allow a lawsuit against the kingdom by victims of the 9/11 attacks, further complicating the U.S.-Saudi relationship.[27]

In Iran, Supreme Leader Ayatollah Ali Khamenei, following the July 2015 announcement of the Joint Plan of Action to constrain Iran's nuclear program, warned the Revolutionary Guard Corps to be ever vigilant against the efforts of the revolution's enemies to "gradually change the people's beliefs." He urged the Corps to take action to counter Western efforts to influence the country's politics and culture.[28] And he ruled out dealings with the United States on issues apart from the nuclear agreement in order to prevent American "infiltration" of Iranian society and the "imposition" of its alien values.

U.S. advocacy of electoral democracy, rule of law, transparency and accountability, freedoms of religion and expression, as well as nondiscrimination on the basis of gender, sexual orientation, or gender identity, is both politically polarizing and highly threatening to many authoritarian governments—as well as to some segments of the U.S. population.

The international environment thus is fragmenting on the basis of conflicting concepts of cultural and religious identity, social values, and political norms. The next administration, as with its predecessors, faces the policy dilemma of building domestic and allied support

---

[27] Karoun Demirjian and Juliet Eilperin, "Congress Overrides Obama's Veto of 9/11 Bill," *Washington Post*, September 28, 2016.

[28] "Iran's Leader Tells IRGC to Guard Against Western Influence," Payvand Iran News, September 17, 2015.

through appeals to shared values as well as interests, while at the same time trying to work with governments threatened by these positions.

Weak States, Misgoverned Areas

Historically, nation-states have mainly worried about other state rivals. This was true of the United States—until 9/11. The Clinton administration in 1994 named five "recalcitrant and outlaw states"—North Korea, Cuba, Iran, Iraq, and Libya.[29] But it was not until 9/11 that weak or misgoverned territories moved to the top of the U.S. security agenda.[30] Places that had been of little import to U.S. interests suddenly mattered once they were viewed as possible safe havens for terrorists plotting attacks. Over the next 14 years, the number of such hot spots on the U.S. national security map exploded as al Qaeda and its progeny took root in new lands.

Threats to U.S. national security may now emanate from strong or authoritarian states that flout international norms, such as Russia or China; brittle states that possess nuclear weapons or the means to develop them, such as North Korea, Pakistan, and Iran; and failing or misgoverned states that cannot or will not enforce the rule of law on their territories, such as Somalia, Syria, and Libya. The recognition that threats that once came from "somewhere" can now come from "anywhere" contributes to the meme of a world that is falling apart.[31]

---

[29] Anthony Lake, "Confronting Backlash States," *Foreign Affairs*, Vol. 73, No. 2, March/April 1994.

[30] The new emphasis on threats from weak states was articulated in the 2002 National Security Strategy, and had bipartisan resonance. For example, Susan E. Rice, before she became national security adviser, worked with Stewart Patrick on such a list. Susan Rice and Stewart Patrick, *Index of State Weakness in the Developing World*, Washington, D.C.: Brookings Institution, 2008. See also, Angel Rabasa et al., *Ungoverned Territories: Understanding and Reducing Terrorism Risks*, Santa Monica, Calif.: RAND Corporation, MG-561-AF, 2007.

[31] Amid the turmoil of 2014, "Is the World Falling Apart?" became a topic of discussion in leading U.S. research institutes and foreign policy publications. See, for example, Thomas Carothers et al., "Is the World Falling Apart?" Carnegie Endowment for International Peace, August 14, 2014. In 2015, Lindsey Graham, the senior senator from South Carolina, said that he was running for President because "I think the world is falling apart." Rebecca Kaplan, *CBS*

Problems Beyond Our Control—Perhaps

> The failure to understand and address risks related to technology, primarily the systemic cascading effects of cyber risks or the breakdown of critical information infrastructure, could have far-reaching consequences for national economies, economic sectors and global enterprises.
>
> World Economic Forum[32]

In a nation that values optimism and a can-do attitude toward any problem, it is politically hazardous to suggest that some international problems are beyond America's ability to solve. Yet the past 15 years have shown the limits of U.S. power and the dangers of overreach. Pessimism is poisonous, but so is the premise that America's mission is to tackle the world's worst problems—or, if it cannot, then it should resist trying. The art of solving a problem is to define it coherently in the first place. Although policymakers tend to divide the world by region and devise strategies accordingly, this approach can falter in the face of multidimensional problems that involve the interaction of regional and global trends. A different and perhaps more useful way of parsing today's complex set of international challenges is to divide them into two pairings of familiar and unfamiliar and enduring and emerging, then tailor expectations, policies, and capabilities to each.

A taxonomy such as the one in Figure 3.2 allows us to separate problems into three categories: familiar challenges that the United States has tackled in the past, with which it can probably cope using existing tools; "hybrid" problems that have both familiar and unfamiliar characteristics, which will likely require new thinking, tactics, or adaptations; and truly new problems (such as rapid climate change and threats to cybersecurity), which are not well understood or for which appropriate or cost-effective strategies, technologies, or resources are likely lacking.

This Morning, May 18, 2015. For a rebuttal, see James Dobbins et al., *Choices for America in a Turbulent World*, Santa Monica, Calif.: RAND Corporation, RR-1114-RC, 2015, pp. 9–12.

32 World Economic Forum, *The Global Risks Report 2016*, Geneva, 11th edition, 2016, p. 18.

Figure 3.2
Familiar and Unfamiliar Challenges

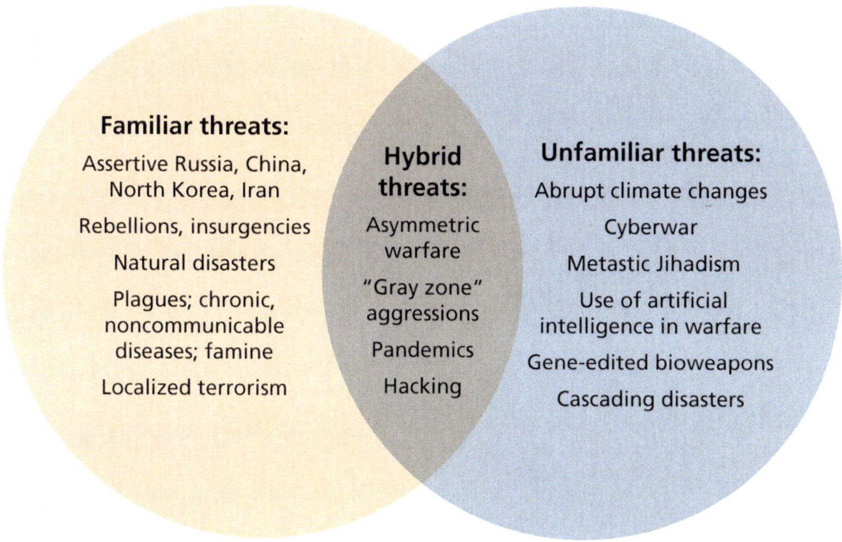

RAND *RR1631-3.2*

Table 3.1 describes these three types of problems in detail. Each type plagues the world today, sometimes in isolation, other times in a specific region, and sometimes coinciding with other problems in time and space.

Familiar Problems

Familiarity does not make problems necessarily easier to solve. They may persist because they are expensive, intractable, bloody, or recurring. Often they arise from states that break the international rules; and therefore they can be addressed by traditional tools of statecraft: diplomacy, defense, deterrence, alliances, economic and military assistance, economic coercion, public diplomacy, subversion, and, as a last resort, war.[33]

The United States faces a familiar problem from states that are trying to challenge the American-led status quo (China and Russia, and,

[33] Dobbins et al., 2015.

Table 3.1
Characteristics of Challenges

Variable	Familiar	Hybrid	Unfamiliar
Characteristics	Problems, however intractable or expensive, that the United States has managed or overcome before Remedies are known, however difficult to apply Often arise from states that break the international rules and therefore can be addressed by traditional tools of statecraft May be law enforcement or public health problems	Emerging problems that have familiar and unfamiliar characteristics that require new thinking or responses May emerge from a concatenation of developments that might be manageable individually but create chaos when one compounds another	United States has little or no experience Deterrence likely does not apply Solutions require extensive international cooperation Remedies are unknown, or existing methods or technologies are unproven or prohibitively costly May combine with other factors to cause cascading failures or compounding problems May also generate unforeseen opportunities
Examples	Great-power competition among states, sometimes leading to warfare.	Hybrid warfare—the blending of regular warfare (identified troops), with irregular warfare (unmarked troops, sabotage, infiltration), making it difficult to identify the adversary	Hybrid warfare + cyberattack
	Major-power nuclear adversaries	Nuclear proliferation to more, smaller states	Proliferation of nuclear materiel or weapons to terrorists or unstable polities.

Table 3.1—Continued

Variable	Familiar	Hybrid	Unfamiliar
Examples (Continued)	Insurgencies within states Rebellion and civil war Religious wars	Transnational insurgencies, revolutionary movements that transcend regions and have international ideological appeal (Spanish civil war, Communist revolution)	State capture by terrorists Infiltration of experienced foreign fighters Self-radicalization
	Espionage Sabotage Physical attacks on critical national infrastructure in wartime	Hacking of government or corporate computers for the purposes of theft, espionage, extortion, or embarrassment	Cyberwar causing physical infrastructure damage, such as blackouts, failure of dams or water supplies, paralysis of financial system Hacking to falsify election results.
	Pollution Natural disasters Conflict over resources Refugee crises	Transnational pollution (acid rain, ozone damage) Greenhouse gas warming and its effects on energy, transportation, water, agriculture, health, and migration	Abrupt climate change advancing with unknown speed and effects, such as rapid sea-level rise[a] or mass migration of climate refugees
	Plagues High rates of non-communicable disease	Pandemics Increased disease due to warmer temperatures	Antibiotic resistance Genetically modified pathogens
	Space race Attacks on ground assets of national space programs	Attacks on satellites or spacecraft in space	Unattributable cyberattack on a satellite

Table 3.1—Continued

Variable	Familiar	Hybrid	Unfamiliar
Available tools	Intelligence, diplomacy, defense, deterrence, alliances, economic and military assistance, economic coercion, public diplomacy, subversion, and as a last resort, war	Coordinated international responses to isolate threats; support for improved governance within affected areas; covert instruments where applicable	No known, affordable, feasible tools International cooperation often necessary to develop and enforce solutions
	Actionable intelligence and prompt action are necessary	Anticipation and resilience are necessary	Resilience is imperative to reduce negative consequences

[a] DeConto and Pollard calculated that the melting of Antarctic ice could raise sea levels by more than a meter by 2100, and more than 15 meters by 2500, if greenhouse gas emissions continue unabated. Robert M. DeConto and David Pollard, "Contribution of Antarctica to Past and Future Sea-Level Rise," *Nature*, Vol. 531, No. 7596, March 31, 2016.

to a lesser extent, North Korea and Iran). Nuclear proliferation is a familiar problem: At the moment, only states have nuclear weapons, and they must be deterred from giving them to other states or to such nonstate groups as terrorists. This national interest has not been altered since 1945.

Nuclear terrorism, however, would be a new problem. If terrorists gained access to a nuclear weapon, the United States and its allies and partners might have only limited capability to detect and disarm it. They have had little practice in mounting large-scale emergency cooperative efforts to avert a terrorist nuclear attack or address the consequences of a detonation.

Hybrid Problems

These might be defined as challenges that have familiar elements but have evolved in ways that make them even more difficult to deal with. Hybrid threats may develop as a result of a concatenation of familiar problems that might be manageable enough individually, but that create chaos when one compounds another.

Hybrid threats will require new approaches, doctrines, training, and potentially new international norms for state behavior. For example, Russia's use of "little green men," or irregular soldiers without insignia, in Ukraine presents particular problems for the United States because it poses a clear threat to U.S. interests without rising to the level where it could be considered an attack under existing rules of engagement.[34]

Hybrid threats are also posed by malevolent groups applying powerful technologies in unforeseen ways. The Obama administration faced a dilemma in responding to North Korea's 2014 cyberattack on Sony Pictures, for example, because a state attack against a corporation, with political but no apparent military purpose, did not meet the legal definition of an act of war. President Obama instead termed it "an act of cyber vandalism"—a relatively new term.[35] Other hybrid threats might be uncontrolled pandemics such as Ebola or Zika virus outbreaks, coastal flooding and water shortages due to climate change, or massive refugee flows due to such events.

Some challenges posed to states by insurgents can be defined as hybrid problems if they have familiar goals (for example, seizing political power) but use novel or asymmetric tactics (improvised explosive devices or the Internet) to achieve them. Terrorist organizations such as the Tamil Tigers in Sri Lanka or the Muslim rebels in Chechnya presented a hybrid problem. While they employed asymmetric or novel techniques, such as the use of female suicide bombers, they had traditional goals—namely, autonomy or resisting foreign occupation. While the suicide bombers themselves could not easily be stopped, these conflicts and terrorist attacks could have been ended through negotiated political solutions.[36]

---

[34] Major John R. Davis, Jr., "Continued Evolution of Hybrid Threats: The Russian Hybrid Threat Construct and the Need for Innovation," *Three Swords Magazine*, Vol. 28, 2015, p. 21.

[35] For a discussion of cyber terminologies, see Brian Fung, "Obama Called the Sony Hack an Act of 'Cyber Vandalism.' He's Right." *Washington Post*, December 22, 2014.

[36] For insightful thinking on this subject, see Audrey Kurth Cronin, "How al-Qaida Ends: The Decline and Demise of Terrorist Groups," *International Security*, Vol. 31, No. 1, Summer 2006; and Cronin, *How Terrorism Ends: Understanding the Decline and Demise of Terrorist Campaigns*, Princeton, N.J.: Princeton University Press, 2009.

In practice, it has often proved difficult to broker peace in such intractable conflicts. The wars in Sri Lanka and Chechnya only ended when the rebels were crushed militarily. Preventive action, however difficult, may be the most promising path to preventing hybrid threats from creating large-scale conflicts.[37]

Pandemics might be thought of as a hybrid problem. Plagues are ancient, but not the ability of viruses to travel by airplane across continents inside asymptomatic hosts. The frequency of infectious disease outbreaks has increased significantly since 1980.[38] As shown by the Ebola outbreak in West Africa and the Zika virus in the Americas, globalization makes it more difficult than ever to stop the spread of infectious diseases. Strengthening public health systems in many of the world's poorest countries and even among low-income communities in the United States is an arduous and expensive task, but one that the U.S. Agency for International Development (USAID), the World Health Organization, and other international authorities have been tackling for some time.

Hybrid problems may be addressed by what the military calls layered *defense in depth*—the U.S. approach to homeland security since 9/11. Terrorist attacks have been thwarted through a suite of measures whose individual efficacy is hard to quantify: hardening targets, improving intelligence and surveillance, better vetting of visa applicants, more-robust border control, better satellite surveillance, and more resources for law enforcement—in short, a hybrid approach. Military strategists are giving increasing attention to "gray zone" operations that do not rise to the threshold of traditional war yet challenge U.S. security interests.[39] For climate change, defense in depth means not only managing the massive transition in energy use, but also preparing for and adapting to the inevitable changes in sea level, storm

---

[37] Paul B. Stares, *The Preventive Imperative: How America Can Avoid War, Stay Strong, and Keep the Peace in the 21st Century*, New York: Colombia University Press, forthcoming, 2017.

[38] Katherine F. Smith et al., "Global Rise in Human Infectious Disease Outbreaks," *Journal of the Royal Society Interface*, Vol. 11, No. 101, October 29, 2014.

[39] See Michael J. Mazarr, "Struggle in the Gray Zone and World Order," *War on the Rocks*, blog, December 22, 2015; and Phillip Kapusta, "The Challenge of Operating in the Gray Zone," *Special Warfare*, Vol. 28, No. 4, October-December 2015.

and precipitation patterns, floods and droughts, and mass migrations that are likely to follow.

Unfamiliar Problems

The world also poses transformative challenges that U.S. policymakers have not encountered before and for which they do not yet have satisfactory solutions. By definition, unfamiliar challenges could have far-reaching consequences but their risks are difficult to assess. Moreover, interactions among the increasing number of variables in the current world equation—and hence the compounding or cascading risk—are nearly impossible to anticipate.[40] Policymakers tend to plan and budget for risks that are to some degree foreseen and understood; efforts are understandably limited when it comes to investigating and preventing low-probability, high-consequence developments for which short-term solutions are deemed unlikely. These emerging and existing problems often do not receive the attention they deserve from senior officials because they are often seen as problems whose solutions—if there are any solutions—lie too far in the future to be worth spending much time thinking about now. In a world of constant technological change, this is a cognitive bias that should be challenged. Ways of doing so are addressed in the Anticipation section in Chapter Five.

Building public consensus for active engagement on unfamiliar problems requires leaders to admit that some problems may, in fact, be beyond our immediate ability to control, and certainly not

---

[40] Former Secretary of State Shultz explained the difficulty of conducting quantitative analysis on low-probability/high-risk events, such as nuclear risk in the civilian and military sectors:

> Accurately analyzing events where we have little data, identifying every variable associated with risk, and the possibility of a single variable that goes dangerously wrong are all factors that complicate risk calculations. . . . It is possible that a single variable could exceed expectations, go dangerously wrong, and simply overwhelm the safety and risk assessment on which those systems were built

as happened at Japan's Fukushima nuclear power complex in 2011. George P. Shultz, *Issues on My Mind: Strategies for the Future*, Stanford, Calif.: Hoover Institution Press, 2013, pp. 364–365.

on our own—a delicate political task. Climate change falls into this category. Even if the world were to stop emitting carbon tomorrow, the Earth's climate would continue to warm.[41] The UN-led climate agreement focuses on voluntary pledges made by countries after past negotiations made clear that the international system has no ability to create legally binding regulations on greenhouse gases.[42] The United States lacks proven, scalable, low-cost technologies for zero-carbon energy or carbon sequestration. There is no clarity about when such technologies might be developed, how affordable they will be, and how easy they will be to disseminate. There is no certainty over how long it might take for climatic effects to become dangerous to the U.S. homeland, or how to quantify that danger. In the face of such uncertainty, the United States faces difficult questions on how best to balance proactive and reactive responses. That the costs and benefits fall on different regions of the country and on different sectors of the economy also constrains bold steps.

The Russian cyberattacks on the Democratic National Committee and state electoral databases is an unfamiliar development that highlights the problem of low-level attacks that can be used to harass, coerce, or sow chaos, yet still fall short of an act of war.[43] Whether these remain tit-for-tat activities or ultimately escalate into wider, more-crippling attacks is far from known. One especially worrisome cyberattack scenario would be a wide-scale U.S. power outage that could cripple a nation that is deeply dependent on electric power and

---

[41] "A significant amount of climate change is already 'baked into' the system, for two undisputed scientific reasons: The oceans take many decades to warm, and many greenhouse gases persist in the atmosphere for centuries. Even if greenhouse gas emissions were to cease tomorrow, the climate would continue to change for several decades to come." Dobbins et al., 2015, p. 70.

[42] As of October 5, 2016, the Paris Agreement was set to take effect in early November after 55 countries representing at least 55 percent of global greenhouse gas emissions ratified it. Justin Worland, "Paris Climate Change Agreement Set to Take Effect After Quick Ratification Process," Time.com, October 5, 2016. Trump has said he would withdraw from the treaty.

[43] President Obama did not term the Russian election hacking a "cyber attack" but rather termed it "aggressive harassment of U.S. officials and cyber operations aimed at the U.S. election." White House, 2016b.

stores information in digital form without adequate backup or redundancy.[44] This is a hard problem, not least because of limitations on the ability to store electricity. Director of National Intelligence James L. Clapper said that the United States does not foresee a "Cyber Armageddon scenario that debilitates the entire U.S. infrastructure" but rather "an ongoing series of low-to-moderate level cyber attacks from a variety of sources over time, which will impose cumulative costs."[45] However, he warned:

> . . . The cyber threat cannot be eliminated; rather, cyber risk must be managed. Moreover, the risk calculus some private sector entities employ does not adequately account for foreign cyber threats or the systemic interdependencies between different critical infrastructure sectors.[46]

Traditional concepts of deterrence may not function well in cyberspace. Effective deterrence requires that one knows the source of the threat, has made known an intention to deny the attacker victory or to invoke costly punishment, and has the capability to do so. It is not clear that those conditions can be met. Clapper articulated the problem as follows:

> [E]ven when a cyber attack can be attributed to a specific actor, the forensic attribution often requires a significant amount of time to complete. Long delays between the cyber attack and determination of attribution likewise reinforce a permissive environment.[47]

Yet where deterrence fails, resilience can help prepare the country for new forms of disruption as well as familiar acts of nature. Good disaster preparedness and policies aimed at building redundant and

---

[44] Ted Koppel, *Lights Out: A Cyberattack, a Nation Unprepared, Surviving the Aftermath*, New York: Crown Publishers, 2015.

[45] James R. Clapper, Director of National Intelligence, "Worldwide Cyber Threats," statement to House Permanent Select Committee on Intelligence, September 10, 2015, p. 1.

[46] Clapper, 2015, p. 2.

[47] Clapper, 2015, p. 3.

hardened systems will be desirable to cope with cyberattack, severe weather events, natural disasters, infectious diseases, nuclear or industrial accidents, or other unforeseen crises. The unfamiliar characteristics of many emerging challenges should increase the need to formulate "low regrets" national strategies—proactive plans to invest in capabilities that benefit Americans even if disaster does not strike. Implementation of such measures may be costly, and certainly requires careful cost-benefit analyses, but they are neither wasteful nor beyond the ability of federal, state, and local governments to tackle.

Terrorism

Terrorism has become sadly familiar, yet in scale, scope, and ferocity, it is the problem that seems most intractable and largely beyond U.S. ability to control without intolerable sacrifice of the civil liberties on which the "American experiment" is premised. Fear that the government will not be able to protect its citizens against terrorist attacks, including from within, prompts some American voters to favor more-defensive, even isolationist, international policies. Many factors make jihadist terrorism particularly unnerving: random attacks in small communities, as well as large cities, by Western citizens or legal residents; the re-infiltration of foreign fighters; self-radicalization via the Internet; attacks that inflict mass casualties and kill hundreds of people at a time by constantly changing means, from box-cutters to underwear bombs to running over pedestrians with trucks.

Yet it is worth remembering that terrorism is not actually new. In the last century, anarchist terrorists conducted numerous bombings in the United States and Europe and successfully assassinated a U.S. president and several crowned heads of Europe, including Austrian Archduke Franz Ferdinand, whose death triggered World War I.[48]

[48] In 1919, anarchists detonated powerful bombs in seven U.S. cities, including Washington, D.C., but failed to kill their targets. In 1920, they succeeded in bombing New York's financial district, killing 38 people and injuring hundreds. President William McKinley was assassinated by an avowed Polish anarchist in 1901. The Immigration Act of 1903 was called the "Anarchist Exclusion Act" because it aimed (unsuccessfully) to keep anarchists and other extremists out of the United States.

The United States and Europe have faced major terrorism challenges since the 1970s, but U.S. attempts to understand and counter violent extremism are in their infancy,[49] as are rehabilitation programs being piloted in such countries as Saudi Arabia. Prison radicalization, identified as a problem in Egypt decades ago, continues to be a serious problem in Europe and beyond. The ISIS "Caliph" Bakr al-Baghdadi emerged as a leader inside the U.S.-run Camp Bucca prison in Iraq, just as his predecessor, Musab al-Zarquawi, the founder of al-Qaeda in Iraq, was radicalized in a Jordanian prison.[50] While many countries have launched a wide range of efforts to counter violent extremism—including preventing radicalization and recruitment, stopping foreign fighters from leaving and returning to host nations, and rehabilitation programs for captured terrorists—evidence about what works and what does not is scarce. Israel has had long experience in dealing with terrorism and still has not succeeded in stopping it. Its current strategy for dealing with the prolonged conflict with Palestinians has been called "mowing the grass"—a term and a policy that many Americans find abhorrent. "Mowing the grass" has been defined by two Israeli scholars as:

> . . . a long-term strategy of attrition designed primarily to debilitate the enemy capabilities. Only after showing much restraint in its military responses does Israel act forcefully to destroy the

[49] Conclusions about success and failure in countering violent extremism (CVE) are mostly anecdotal, and metrics are scarce. For a useful review of evidence-based evaluations of CVE efforts, see Peter Romaniuk, *Does CVE Work? Lessons Learned from the Global Effort to Counter Violent Extremism*, Global Center on Cooperative Security, September 2015. A recent meta-analysis of 107 reports on CVE found that most studies do not provide program evaluation data, and only 24 "provided data that could be broadly categorized as correlational findings of program effectiveness." Caitlin Mastroe and Susan Szmania, *Surveying CVE Metrics in Prevention, Disengagement and De-Radicalization Programs*, College Park, Md.: National Consortium for the Study of Terrorism and Responses to Terrorism, Report to the Office of University Programs, Science and Technology Directorate, Department of Homeland Security, 2016.

[50] Evan Blackwell, *Prison and Jihad: A Threat Behind Bars*, NATO Association of Canada, October 10, 2015.

> capabilities of its foes as much as possible, hoping that occasional large-scale operations also have a temporary deterrent effect. . .[51]

This is the outcome that the United States seeks to avoid: protracted conflict requiring the unending use of violence in an unwinnable war of attrition that breeds callousness and hatred on both sides. Wars of attrition have resulted in the use of female shooters and suicide bombers; the radicalization of some U.S. citizens and demonization of Muslims by others; and above all, the deliberate incitement of hatred. These tactics are especially corrosive to Western values—which is, of course, their intention.

Many Americans fear that the struggle against terrorism will become a "perpetual war" that will corrode the United States from within. As President Obama put it:

> We cannot use force everywhere that a radical ideology takes root; and in the absence of a strategy that reduces the wellspring of extremism, a perpetual war—through drones or Special Forces or troop deployments—will prove self-defeating, and alter our country in troubling ways.[52]

Yet the President has been forced to continue U.S. military involvement in Iraq, Afghanistan, and then Syria, steadily increasing since the modest re-escalation of U.S. military involvement in Iraq in 2014. Followers of radical Islam are now on the march across four continents. In 2016 alone, Salafi-jihadist terrorists and followers have attacked targets from Brussels to Bamako to Orlando, as well as in Ankara, Baghdad, Berlin, Côte d'Ivoire, Tunis, Jakarta, Iskandariya, Istanbul, Kabul, Lahore, Ougadougou, Mogadishu, Nice, Peshawar, Stavropol, San Bernardino, Tunis, and Zliten.

In March 2016, Secretary of State John Kerry declared that ISIS had committed genocide against Yazidis, Christians, and Shiites in

---

[51] Efraim Inbar and Eitan Sharim, *Mowing the Grass in Gaza*, Begin-Sadat Center for Strategic Studies, Perspective Paper No. 255, July 20, 2014.

[52] Barack Obama, remarks at the National Defense University, Washington, D.C., May 23, 2013.

areas under its control.[53] He stated that ISIS had committed crimes against humanity, ethnic cleansing, and enslaved and raped thousands of women, including Yazidis, Christians, and Shia Turkmen. Congress, with rare unanimity, passed a resolution accusing ISIS of genocide and a second resolution charging the government of Syria and its allies of war crimes and crimes against humanity.[54] While of symbolic importance, these legal designations had no noticeable effect.

However, if and when ISIS is defeated, some fear that strife among the Sunnis, Shiites, and Kurds may continue in the areas that ISIS controlled and beyond as Iran tries to create a buffer zone in Shiite-dominated territory in Iraq, the Kurds push for statehood, and radical Sunni fighters from non-ISIS groups fight Shiite domination. Moreover, the assumption that, without territory, ISIS will no longer be able to recruit, inspire, or assist attackers inside Western countries is unproven.

Potentially Existential Threats: Nuclear Weapons, Biological Threats, Climate Change

Whatever else is on the next president's agenda, U.S. leadership will be required to address three critical threats to the United States and the planet: nuclear weapons, emerging biological threats, and rapid climate change. Each of these problems has the potential to change or even destroy life as we know it. Despite somber officials warning about the dangers of nuclear weapons and bioterrorism, these grim issues rarely rise to the forefront of public debate. Nevertheless, reducing these supranational threats is a core U.S. interest that must remain a constant focus of the government.

As of January 2016, nine nations possessed more than 15,000 nuclear warheads, as shown in Table 3.2.

While the numbers of U.S. and Russian weapons and the likelihood of a nuclear exchange are far less than during the Cold War, today's nuclear issues are more complex. China, Russia, and the United

---

[53] John F. Kerry, "Remarks on Daesh and Genocide," U.S. Department of State, Washington, D.C., March 17, 2016b.

[54] House Concurrent Resolution 75, ISIL atrocities, passed 393–0, March 14, 2016; House Concurrent Resolution, Syria violations, passed 392–3, March 14, 2016.

Table 3.2
World Nuclear Forces, 2016

Country	Year of First Nuclear Test	Deployed Warheads[a]	Other Warheads	Total 2016
United States	1945	1,930	5,070	7,000
Russia	1949	1,790	5,500	7,290
United Kingdom	1952	120	95	215
France	1960	280	20	300
China	1964		260	260
India	1974		100–120	100–120
Pakistan	1998		110–130	110–130
Israel			80	80
North Korea	2006		10	10
Total		4,120	11,235–11,275	15,355–15,395

SOURCE: Stockholm International Peace Research Institute, "Global Nuclear Weapons: Downsizing but Modernizing," web page, June 13, 2016.

NOTE: All estimates are approximate and are as of January 2016. Totals do not include figures for North Korea.

[a] Deployed means warheads placed on missiles or located on bases with operational forces.

States are modernizing their nuclear arsenals, and 2,000 nuclear weapons are on alert worldwide. North Korea's efforts to develop more-capable warheads and delivery systems are advancing, and Iran's ten-year commitment to suspend its nuclear program under the 2015 Joint Comprehensive Action Plan remains under scrutiny.

That terrorists have not succeeded in accessing nuclear material is an underappreciated triumph for the world's governments. Yet there is no room for complacency in light of reports that ISIS and other groups are trying to get material for radiological bombs and the 2,000 metric tons of "nuclear weapons useable material"—enriched uranium and separated plutonium—known to remain in civilian and military facilities around the world.[55] While most of this highly radioactive material would be difficult for terrorists to steal—anyone who tried would

[55] Office of the Press Secretary, "Fact Sheet: The Nuclear Security Summits: Securing the World from Nuclear Terrorism," The White House, Washington, D.C., March 29, 2016.

quickly be killed by the radiation—the possibility of terrorists using less-radioactive materials to detonate a "dirty bomb" in a city has been widely discussed. Such a device would not be likely to kill people or cause immediate injury, but it might do considerable financial and psychological damage by creating panic, triggering evacuations, and putting key urban areas off limits.[56]

International norms aimed at preventing nuclear diversion and smuggling, reducing the sources of material that can be weaponized, and improving international law enforcement cooperation have been greatly expanded and strengthened, and research is under way to find alternative technologies to replace dangerous radioactive sources.[57] The United States, which has been in the vanguard of such multinational efforts for nearly three decades, cannot afford to disengage.

As tensions rise between the United States and Russia, a more dangerous possibility is an inadvertent escalation into nuclear war due to misunderstanding of intentions, miscalculation, or error. As Anthony Barrett writes:

> The use of a single nuclear missile in a populated area would be devastating; the use of substantial fractions of U.S. and Russian nuclear arsenals could trigger a global catastrophe . . . [Yet] at present, the United States does not appear to have a consistently used method for assessing the risk of inadvertent nuclear war.[58]

Of course, the world has no experience of war between nuclear powers, and war of any type has the potential to become nuclear. However, *accidental* nuclear war is eminently preventable. Some wars are planned and intentional, but history shows that most are the result of leadership error. This has included

> gross overconfidence, sloppy analysis, lapses of objectivity, wrongheaded preconceptions, disregard of facts that cast doubt on those

---

[56] Gregory S. Jones, "ISIS and Dirty Bombs," *The Cipher*, June 3, 2016.

[57] Office of the Press Secretary, 2016.

[58] Anthony Barrett, *False Alarms, True Dangers? Current and Future Risks of Inadvertent U.S.-Russian Nuclear War*, Santa Monica, Calif.: RAND Corporation, PE-191-TSF, 2016.

> preconceptions, zeal at the cost of rationality, suppression of debate, punishment of dissent, and other failures, fallacies, and fantasies.[59]

During the post–Cold War period, the United States, Russia, and, to a lesser extent, China engaged in extensive lower-level bureaucratic engagement aimed at building deep ties, reducing the risk of misunderstanding, and building trust. Military exchanges, so-called "Track II" unofficial consultations among senior figures, academic intercourse, civil society engagement, and personal friendships flourished. To the extent that worsening U.S. relations with Russia and China inhibit these rich and diverse sources of communication and mutual reassurance and leave national leaders more isolated, the risk of blundering into war increases. Needless to say, avoiding nuclear war outweighs almost all other U.S. interests.

Emerging Biological Threats

The Biological Weapons Convention (BWC), which entered into force in 1975, was one of the major accomplishments of the postwar liberal order. It was the first multilateral treaty to ban the development, production, and stockpiling of an entire category of weapons and the means to deliver them.[60] It cemented and expanded the international norm, which had been developing since the 1925 Geneva Conventions, that the use of biological weapons can never be justifiable. As UN Representative Angela Kane put it on the 40th anniversary of the Conventions:

> How many States today boast that they are "biological weapon States"? Who argues now that the bubonic plague and smallpox

[59] David C. Gompert et al., *Blinders, Blunders, and Wars: What America and China Can Learn*, Santa Monica, Calif.: RAND Corporation, RR-768-RC, 2014, p. iv.

[60] United Nations Regional Centre for Peace and Disarmament in Asia and the Pacific, "Biological Weapons Convention Meeting of Experts to Convene in Geneva from 10 to 14 August 2015," web page, August 7, 2015.

> are legitimate weapons to use under any circumstances? Who speaks of a bioweapon umbrella?[61]

At the time of the negotiation of the BWC, the threat from biological weapons was thought to be from states. Therefore the Convention sought to limit any activities dealing with biological material that had "no justification for prophylactic, protective or other peaceful purposes."[62] Today, with advances in the biological sciences and proliferation of technology, capabilities for manipulating biological material have been democratized, allowing access to these technologies to even untrained people. The range of potential biological threats has thereby been expanded to include not just state threats and naturally occurring disease, but bioterrorism, dangerous experiments, and accidents. Traditional biological agents such as weaponized bubonic plague or anthrax remain threats, but now the capacity to manipulate biological agents to increase virulence and transmissibility, create new pathogens from base proteins, and even alter the essence of life by the manipulation of the genome, have become widely available. In short, biotechnology has evolved from an art and science to an industrial and engineering process.

Many analysts warn that the leading biological weapons threat today arises from a three-year-old gene-editing tool called CRISPR. Developed at the University of California, Berkeley, and the Massachusetts Institute of Technology, CRISPR is "a method of snipping and editing genes that bacteria have used for billions of years, but humans have deployed only for about thirty-six months."[63] In February 2016, Director of National Intelligence Clapper identified genome-editing technology as a potential weapon of mass destruction.[64]

---

[61] Angela Kane, UN High Representative for Disarmament Affairs, video message to the Biological Weapons Convention: 40th Anniversary Event, Geneva, Switzerland, March 30, 2015.

[62] Biological and Toxin Weapons Convention Website, University of Bradford, undated.

[63] Laurie Garrett, *CRISPR: Transformative and Troubling*, Council on Foreign Relations, April 13, 2016.

[64] James R. Clapper, Director of National Intelligence, "Worldwide Threat Assessment of the U.S. Intelligence Community," statement for the record to the Senate Select Committee

Long before his remarks, the technology had proliferated to laboratories around the world and was so inexpensive that amateur biologists were using it in neighborhood labs.[65] A Chinese team announced in 2014 that it had used CRISPR to genetically alter monkeys, who then passed the altered traits to their offspring. The technology holds the promise of being able to alter the human genome to prevent genetic blindness, cure HIV, or create drought-resistant crops, but experts fear that it is so easy to use that terrorists or others who do not accept voluntary ethics guidelines could soon be able to engineer powerful new diseases that could devastate humans or creatures.[66] By 2017, CRISPR kits were available online for as little as $450.[67]

The United States has asked scientists to comply with ethics guidelines barring the experimentation of gene-editing on humans, but Clapper noted:

> Research in genome editing conducted by countries with different regulatory or ethical standards than those of Western countries probably increase the risk of the creation of potentially harmful biological agents or products. Given the broad distribution, low cost and accelerated pace of development of this dual-use technology, its deliberate or intentional misuse might lead to far-reaching economic and national security implications.[68]

---

on Intelligence, February 9, 2016.

[65] Daniel M. Gerstein, "How Genetic Editing Became a National Security Threat," *Bulletin of the Atomic Scientists*, April 25, 2016.

[66] Gerstein, 2016.

[67] For example, the OriGene CRISPR Cas9 Starter Kit is sold by OriGene Technologies. Less expensive versions for home experimentation are also available. See Loz Blain, "Do-It-Yourself CRISPR Genome Editing Kits Bring Genetic Engineering to Your Kitchen Bench," Gizmag.com, November 12, 2015; OriGene Technologies, "CRISPR Cas9 Starter Kit," web page, undated.

[68] Clapper, 2016.

One month later, CRISPR was used to create a living organism that includes the minimal genome considered necessary for life[69]—a synthetic new species that doubles in population every three hours—inside a laboratory in the J. Craig Venter Institute in La Jolla, California.

Public health researcher Laurie Garrett described it as "a completely novel life form . . . unlike anything that exists in nature: It is alive and can self-reproduce, passing its genes on in a totally new stream of evolution."[70]

Along with nuclear weapons and climate change, the development of genetically enhanced pathogens or synthetic life forms and their potential uses raise existential questions.

For the immediate purposes of national strategy, future administrations will need to focus on bioweapons, bioterror, and bio-error, including the accidental release of genetically altered material. It should consider leading an effort to update the Biological Weapons Convention, UN Security Council resolutions that govern the proliferation of weapons of mass destruction, and/or other national laws and standards that might be adapted to regulate genetic engineering. Policymakers will also need to rethink how the world could meaningfully enforce such rules, assuming states could agree on them.

Climate Change

Scientists have amassed overwhelming evidence that rising greenhouse gas concentrations are already changing the Earth's climate, increasing the incidence of extreme temperatures, exacerbating drought, raising sea levels, acidifying oceans, disrupting agriculture, and increasing the intensity of storms. Figure 3.3 charts the historical trends in greenhouse gases.

---

[69] Clyde A. Hutchison III et al., "Design and Synthesis of a Minimal Bacterial Genome," *Science*, Vol. 351, No. 6280, March 25, 2016.

[70] Garrett, 2016. See also, Ewan Callaway, "'Minimal' Cell Raises Stakes in Race to Harness Synthetic Life," *Nature News*, Vol. 531, No. 7596, March 26, 2016.

Figure 3.3
Global Atmospheric Green House Gas Concentration

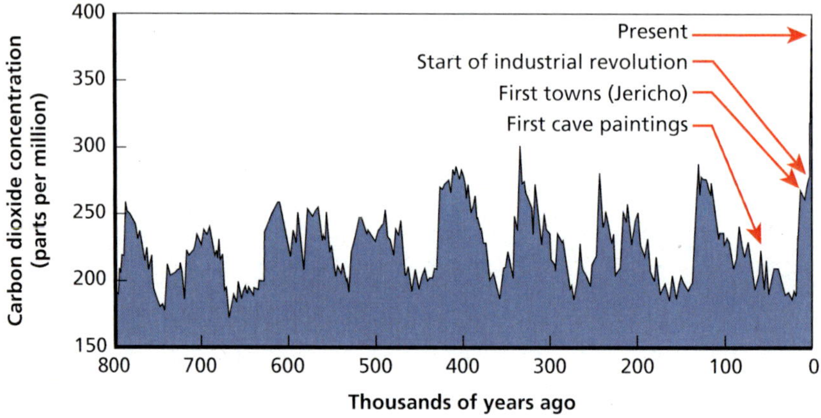

SOURCE: Dobbins et al., 2015, Figure 6.1, based on data from Andrew Freedman, "The Last Time CO2 Was This High, Humans Didn't Exist," *Climate Central*, May 3, 2013.
NOTE: Greenhouse gas concentrations are higher than they have been in nearly a million years and are rising faster than they have in the entirety of human civilization.
RAND *RR1631-3.3*

U.S. public opinion on climate change has shifted dramatically,[71] but many Americans and some political leaders have been significantly more skeptical about climate change than other publics.[72] There is no domestic political consensus about how much, if anything, the United

[71] "Global Warming," CBS News/New York Times Poll, November 18–22, 2015, p. 2. Out of 1,030 surveyed, the percentage of those saying global warming is causing an impact now rose from 35 percent in 2001 to 50 percent in 2015, while those saying it will have an effect in the future fell from 41 percent in 2001 to 25 percent in 2015. The percentage who said climate change would have no serious impact rose from 17 percent to 19 percent, while 1 percent said global warming did not exist at all. A 2016 Gallup survey found 64 percent said they worried a "great deal" or a "fair amount" about climate change, the highest level in eight years. "U.S. Concern About Global Warming at Eight-Year High," Gallup, web page, March 16, 2016.

[72] The publics in the United States and the Middle East generally see the climate change issue as less serious, compared with those in Latin America, Africa, and Europe. See "Spring 2015 Global Attitudes Survey," Pew Research Center, web page, Washington, D.C., June 23, 2015, questions 32, 41, and 42. Nevertheless, 66 percent of Americans surveyed by Gallup at the time of the 2015 Paris UN conference on climate change supported the United States joining an international treaty to limit greenhouse gas emissions, while only 27 percent said the United States should not join. "Global Warming," 2015.

States should spend to mitigate climate change, and President-elect Trump has vowed to cancel billions in UN climate change spending,[73] causing friction with U.S. allies.[74]

Achieving consensus will be difficult because the costs would fall more heavily on some parts of the country and economy—for example, those regions whose economies are tied to coal—than on others. Consensus will also be complicated by multiple great unknowns: the costs of decarbonizing the world economy—that is, of reducing or eliminating net human greenhouse gas emissions; the extent of the damage that will occur if climate change is not controlled; and the timing both of climate change itself and of the human activities to halt it or adapt.

This vast range of uncertainty stems both from the novelty of the science—for millions of years, the Earth has not been as warm as it may become, so any estimate of how the climate will behave is an extrapolation of the previously unobserved—and the difficulty of predicting how different societies and biological systems will respond to the changes. For instance, the potential for technological innovation is unknown—how fast will entrepreneurs lower the cost of solar power and electric cars? How effective will various government policies aimed at incentivizing decarbonization turn out to be? How quickly will people in different countries change their lifestyles? The three strategic concepts outlined in Part II of this report in effect lay different bets not only on the severity of the changes in climate that will occur but also on how different countries will react. Each of these bets carries downside risks; indeed, for one or more of these possible bets, the outcome could be catastrophic. Thus, hedging strategies will be a necessity under any path.

As a strategic problem, climate change poses serious national security challenges under some sets of assumptions, because many of the areas projected to be hardest-hit by climate change are also the countries that tend to be poorest, least stable, and receiving billions in U.S.

---

[73] Trump vowed to "use the money to fix America's water and environmental infrastructure." Donald J. Trump, "Donald Trump's Contract with the American Voter," web page, November 10, 2016.

[74] Nina Chestney and Alister Doyle, "France, U.N. Tell Trump Action on Climate Change Unstoppable," Reuters, November 15, 2016.

foreign economic and/or military assistance. A map focused on rivers and coastal areas likely to be affected most by climate change is shown in Figure 3.4. Areas already rife with conflict in Africa, South Asia, and Southeast Asia show the greatest vulnerabilities. Climate change will likely be a multiplier of stress and instability in these and other regions.

As a diplomatic problem, regardless of its ideology, the next administration can expect to face continuing pressure to give more international climate aid, both because the poorest countries are least responsible for high emissions but often most vulnerable to a shifting climate, and because the United States may have military, counterterrorism, or other important interests in precisely these regions.

Above all, however, climate change poses an enormous challenge to policymakers required to make irreversible decisions amid conditions of deep uncertainty. There is a huge range in the alternative scenarios for global temperature increase and an even more staggering range of estimates about how much damage might ensue. All of these factors make climate change a quintessentially difficult challenge of risk management.

A recent meta-analysis provides a concise summary of the wide range of cost projections contained in the 2015 Intergovernmental Panel on Climate Change Fifth Assessment Report.[75] It estimated that action to hold the increase in global temperature below 2 degrees Celsius (C) could cost anywhere from a few tenths of a percent to more than 10 percent of gross world product (GWP) per capita by 2100. For perspective, the growth rate of GWP over the last ten years for developed economies has been as high as 3.1 percent in 2010 and as low as –3.4 percent in 2009.[76] Even excluding other factors, the World Bank has estimated that water scarcity, exacerbated by climate change,

---

[75] Laurent Drouet, Valentina Bosetti, and Massimo Tavoni, "Selection of Climate Policies Under the Uncertainties in the Fifth Assessment Report of the IPCC," *Nature Climate Change*, Vol. 5, February 2015.

[76] International Monetary Fund, World Economic Outlook Database, October 2016b.

Figure 3.4
Map of Climate Change Hot Spots

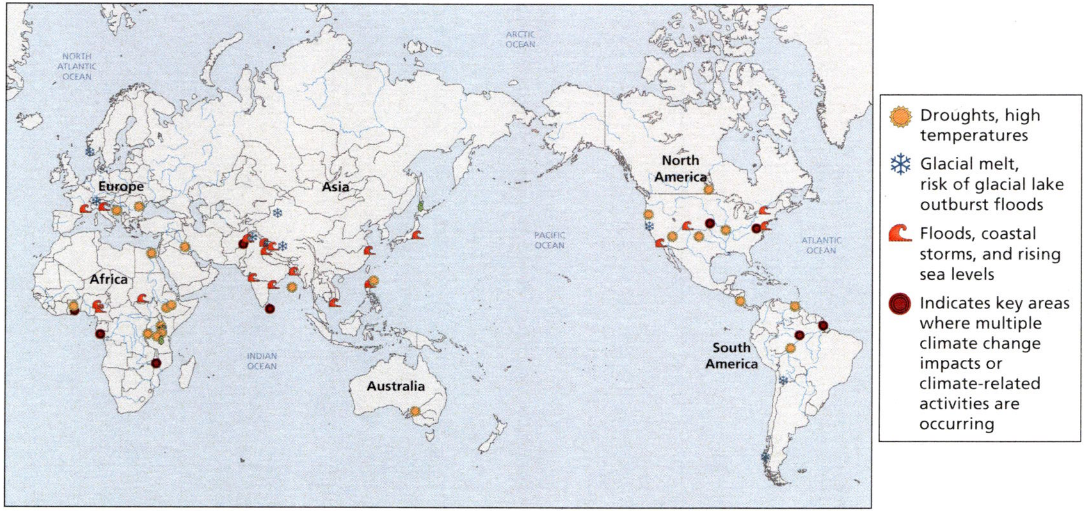

SOURCE: International Rivers, "Map of Climate Change Hotspots," (CC BY-NC-SA 3.0 US) with author overlay, undated.

RAND *RR1631-3.4*

Table 3.3
Possible Scenarios for Climate Change Impacts

	Cost of Decarbonizing Global Economy	
Impacts of Climate Change	**Inexpensive (< 1% GWP)**	**Expensive (5 to 10% GWP)**
Large reduction in GWP in 2100 (5 to 40%)	"Easy to be green" Decarbonize rapidly to avoid worst climate effects and reap co-benefits such as reductions in air pollution.	Worst outcome Accept significant economic costs to decarbonize, or accept significant domestic adaptation costs and significant global instability, refugee flows, conflicts over resources, and, possibly, wars.
Small reduction in GWP in 2100 (< 1%)	Best outcome: Manage climate change through gradual adaptation and shift to efficient energy technologies.	"Live with climate change" Rapid decarbonization without more-affordable technologies would impose opportunity costs. Let other nations lead, help the poorest adapt, focus on domestic adaptation.

will reduce GDP in some regions by up to 6 percent by 2050, without active steps to manage water resources.[77]

However, the meta-analysis also found the potential range of costs of not taking prompt steps to reduce emissions could be far greater. It found the cost of damage that could result from unabated greenhouse gas emissions could range from a few tenths of one percent to more than 40 percent of GWP per capita by century's end.

Table 3.3 offers four alternative scenarios that match climate change costs and damages to policy outcomes. If the impacts of unlimited climate change prove relatively small, or if the cost of decarbonizing the global economy proves relatively inexpensive (both less than roughly 1 percent of GWP per capita), then climate change will impose new demands but none beyond the scope of technological change that humans have handled in the past. If the impacts of unmitigated climate change are large and decarbonization proves expensive (both greater than roughly 5 percent of GWP), then climate change will place unprecedented demands on governments and societies.

---

[77] Even with efficient water policies in place, the study estimated that the Middle East and North Africa would still face GDP drops of 6 percent, whereas better water management would accelerate China's GDP by 2 percent and Central Asia's by 6 percent by 2050. See World Bank, *High and Dry: Climate Change, Water and the Economy*, Washington, D.C., May 2016c.

CHAPTER FOUR

Leveraging U.S. Strengths, Dealing with Vulnerabilities

Over the past quarter-century, the advanced industrial economies have been struggling to adapt to three large historic trends: First, the entry of 3.5 billion new people into a globalizing economy; second, rapid technological advances that have created and destroyed whole industries; and third, the shifting demographics of higher-income societies (with the same shift soon to affect middle-income and low-income economies). Each of these transformations is expected to accelerate in the decades to come. Each will require short-term domestic policy adjustments in order to bend the long-term trend lines in the United States' favor.

The United States possesses enormous strengths and competitive advantages that have enabled it to thrive for more than a century in the face of determined adversaries. Chief among them have been the adaptability of the nation and its citizens to confront daunting challenges, domestic and international; its culture of innovation; and its ability to garner friends and partners across the globe.

At the same time, the country today has structural economic weaknesses, as well as deep political and cultural divisions that were sharpened by the 2016 election. If not addressed, the political and fiscal ramifications of these vulnerabilities may constrain America's ability to mount an effective, coherent foreign policy.

A successful international strategy should therefore recognize and reinforce the country's significant strengths in adapting to an innovation-fueled, automation-enabled service economy, and minimize the associated socioeconomic vulnerabilities. This chapter addresses these strengths along with the domestic policy concerns that are most relevant to securing the U.S. position in an increasingly competitive world.

U.S. Strengths

Despite a 2016 election that focused attention on U.S. weaknesses and failures, America's economic position vis-à-vis its main competitors is strong. The protracted and anemic U.S. economic recovery appears to have taken root at last, albeit leaving pockets of the country behind. Real median household income increased by 5.2 percent between 2014 and 2015, to $56,516.[1] The U.S. unemployment rate, at less than 5 percent,[2] is low compared with most other advanced economies and considered by many economists to be near full employment.[3] New business startups, an engine of job creation, leaped from 2014 to 2015 after declining since 2010.[4]

Debt, jobs, dislocation, and demographics are problems for many advanced economies, yet the United States is better positioned than any of its rivals to address them. Its first advantage is the power of agency. The reforms needed to sustain U.S. economic vitality are within America's power to enact. This is not the case for many other countries, which are more subject to external pressures—the strength of the global economy, the changes in market demand and supply conditions, and their levels of debt, among other factors.

A second U.S. competitive advantage is the vibrant and innovative private sector, which has generally good access to capital and decades of experience in competing in global markets. Although segments of the

---

[1] U.S. Census Bureau, "Income, Poverty and Health Insurance Coverage in the United States: 2015," News Release CB16-158, September 13, 2016b.

[2] Unemployment has hovered at 5 percent or less between October 2015 and October 2016. U.S. Bureau of Labor Statistics, "The Employment Situation, October 2016," news release, November 4, 2016.

[3] Of the OECD countries, the United States in 2015 had lower unemployment rates than all but the United Kingdom (tied with the United States at 5.3 percent), the Czech Republic, Germany, Iceland, Japan, Mexico, Norway, South Korea, and Switzerland. OECD, "Short-Term Labour Market Statistics: Harmonized Unemployment Rate," database, undated-c.

[4] Robert W. Fairlie et al., *The Kauffman Index: Startup Activity: National Trends*, Kansas City, Mo.: Ewing Marion Kauffman Foundation, 2015.

American public are increasingly skeptical of globalization,[5] the United States benefits from its openness to the global economy in a number of ways. Trade brings expanded markets for U.S. goods, enabling firms to reach greater economies of scale. Imports bring cheaper inputs and cheaper goods. The United States has an advantage in services trade, and the globalization of services expands high-skill and high-compensation employment. Openness to investment brings in new capital, as well as new management knowledge and ways of doing business. Likewise, the ability to invest abroad allows U.S. companies to better meet foreign market demand at lower cost, to localize output more easily. Global use of the dollar has enabled the U.S. government to borrow at lower costs than otherwise and to borrow in dollars, avoiding the problem of exchange rate risk that besets many other countries.

The United States also benefits from its leadership of the global financial institutions. Even as China attempts to build institutions to challenge U.S. financial dominance, such as the Asian Infrastructure Investment Bank (AIIB) or the New Development Bank,[6] it models them after U.S.-built structures and keeps a significant portion of its reserves in U.S. government debt instruments—more than $1.2 trillion in March 2016. As one economist noted, from 2003 through 2007, foreigners poured $7.8 trillion of new foreign investment into the United States—more than $5 billion a day—and "even after the subprime crisis started to unfold, the money still kept 'rolling in' to the United States, albeit at a slower pace."[7] Capital continues to leave China after its stock market turmoil of 2015–2016. Indeed, concern over capital flight may limit Chinese leaders' willingness to further devalue its currency.

---

[5] In an April 2016 survey of 2008 respondents, 49 percent, including 55 percent of Republican voters, said U.S. involvement in the global economy is "a bad thing as it lowers wages and costs jobs." Only 44 percent saw it as "a good thing, because it provides the U.S. with new markets and opportunities for growth." Pew Research Center, 2016b, p. 19.

[6] The AIIB is to be capitalized at $100 billion, with China contributing the single largest share.

[7] Mihir A. Desai, C. Fritz Foley, and Kristin Forbes, "Financial Constraints and Growth: Multinational and Local Firm Responses to Currency Depreciations," *Review of Financial Studies*, Vol. 21, No. 6, November 2008.

In the mid-2000s, there was great concern that the rising U.S. trade deficit would damage the economy.[8] The current account deficit hit almost 6 percent of GDP in 2006, raising fears that the United States was living beyond its means on foreign financing, and that foreigners could tire of accumulating dollars, causing dollar depreciation. Instead, the current account deficit had narrowed to 2.2 percent of GDP by 2014, helped by a surplus in services trade.[9] Increases in exports of services are an indicator that the global market sees value in the U.S. "knowledge economy"—a bright spot as the country seeks to boost its advantage in a competitive world.

U.S. strengths are also reflected in most global growth projections, which indicate that U.S. growth prospects are better than those of any other major developed country. Even among the world's top five developing nations—Brazil, Russia, India, China, and South Africa—U.S. prospects outstrip those of three: Brazil, Russia, and South Africa.[10] In the public mind, discussion of China's extraordinarily rapid growth may overshadow the large advantages in wealth, health, and education that the United States has accumulated over seven decades of growth since World War II. In 2014, U.S. GDP was $17.4 trillion, more than the total GDPs of Japan, Germany, the United Kingdom, France, and Brazil combined.[11] China's was $10.4 trillion. In terms of purchasing power parity, the United States was the ninth wealthiest country in the world in 2015, at $57,045 per capita.[12] (Qatar ranked first at $146,000 per capita).[13] The OECD, which has been developing alternative methodologies to GDP for measuring national well-

---

[8] Paul Krugman, "CSI: Trade Deficit," *New York Times*, April 24, 2006.

[9] Shatz, 2016, p. 29.

[10] Shatz, 2016.

[11] World Bank, "GDP Ranking," web page, April 11, 2016b.

[12] These 2015 rankings calculated GDP based on purchasing power parity. Valentina Pasquali, "The World's Richest and Poorest Countries," *Global Finance*, web page, November 1, 2015.

[13] However, because living costs are lower there, China surpassed the United States in 2014 using the purchasing power parity price-adjusted comparison, because U.S. GDP at market rates was $17.4 billion, whereas China's was $18 billion. Shatz, 2016, p. 46.

being, calculated that the United States is better off than average in the 34 member countries in terms of average household net-adjusted disposable income per capita ($41,071 per year), secondary education (90 percent of adults), air pollution levels (low), and overall life satisfaction (slightly better than average).[14] Despite political opposition to immigration, immigrants, skilled and unskilled, are an economic asset to the United States and have long contributed to the culture of entrepreneurship. More than a quarter of U.S. startups in 1997 were founded by immigrants.[15] Immigration has also helped the United States maintain vitality through its population size.[16]

The United States also continues to enjoy enormous comparative political strengths after seven decades of international leadership.[17] Despite its domestic political fluctuations, it has been able to maintain policy continuity in its international economic positions, which has lent the United States credibility. Whether this advantage can be sustained is unclear.

Finally, the United States has cultural and institutional strengths that have been both commercially and politically valuable. It is the world's leading exporter of movies, books, and music.[18] A recent Chinese army recruiting video was set to hip-hop music, an American cul-

---

[14] OECD, "Better Life Index," web page, undated-a. China is not a member of the OECD, but it, too, has attempted in recent years to measure citizen well-being beyond GDP, publishing its own indexes of "green development" and poverty reduction.

[15] Immigrant entrepreneurs now account for 28.5 percent of new entrepreneurs in the United States, compared with 13.3 percent in 1997. Fairlie et al., 2015, p. 6.

[16] Immigrants constituted 34.7 percent of U.S. population growth between 1990 and 2000, 33.6 percent between 2000 and 2010, and 23.5 percent between 2010 and 2014. Because economic growth rates are the result of growth in capital, labor, and productivity, sustained and significant levels of immigration will help the United States retain the standing and influence in the world that stems from its economic size. Shatz, 2016, p. 50.

[17] This argument was made persuasively in Jones, 2014.

[18] The United States has been running a growing trade surplus in exports of arts and culture, estimated at $24.1 billion in 2013. See National Endowment for the Arts, "The Arts and Economic Growth," press release, February 16, 2016.

tural export.[19] Even the new People's Liberation Army uniforms appear to have been inspired by the U.S. digital camouflage look.

The United States hosts more of the top global universities than any other nation.[20] Many global leaders in politics and business have studied in the United States (along with a smaller fraction of U.S. leaders who have studied abroad), and, more often than not, their views and sympathies are shaped by this experience, contributing to a commonality of attitude about norms and values in what is now called "the international community." While none of these advantages is guaranteed to continue, no other nation is equipped to take over the role of global steward or is likely to be considered as benign as the United States in filling that role.

The United States is still the destination of choice for people and their money. It is the still the partner of choice for commercial and military alliances. The rule of law is one of the most powerful of U.S. assets. The world's economic and political refugees are not running east in hopes of a better life in Russia, China, Iran, or North Korea; they are voting with their feet (in destabilizing numbers) for the EU and the United States. The U.S. embrace of diversity has made it possible for the nation to attract and retain the best minds from around the world.[21] One metric is telling: Nearly one-tenth of U.S. billionaires are foreign born. The 2014 Forbes list of the 492 American billionaires

---

[19] "Battle of the Declaration" (Chinese PLA recruiting video), Youku Tudou, website, 2016. See also Andrew S. Erickson, "PLA Recruiting Ad: The Rap Video—Annotated Translations by Dr. Kevin Jensen and Professor Mark Metcalf," Andrew S. Erickson: China Analysis from Original Sources, blog, May 11, 2016.

[20] The United States hosts 13 or 14 of the top 20 universities in various rankings, although its share of international students has dropped—attributed in part to the increase in cost and difficulty of obtaining student visas after 9/11. "World University Rankings, 2015–16," Times Higher Education, London, undated; Jason Lane, "U.S. Losing Its Dominance in Global Higher Education Market," *The Conversation*, October 14, 2015.

[21] Some argue that it is precisely the displacement and marginalization experienced by immigrants that explain their creativity. "Uprooted from the familiar, they see the world at an angle, and this fresh perspective enables them to surpass the merely talented." Eric Weiner, "The Secret of Immigrant Genius," *Wall Street Journal*, January 15, 2016.

includes 50 individuals born in 26 other nations.[22] They include Sergey Brin (born in Russia), Elon Musk (South Africa) John Kapoor (India), Shahid Khan (Pakistan), and Jan Koum (Ukraine).

In sum, the United States is only "in decline" if one assumes that its power is diminished by living in a world of many other rich and successful nations.[23] The declinist view ignores the ability of the United States to benefit from the growth, innovations, and contributions of other nations, including by adopting and adapting knowledge, technologies, and practices developed elsewhere. Nevertheless, during the 2016 campaign, the declinist view strongly resonated with sizable segments of the U.S. electorate that view internationalism, multiculturalism, and globalization to be illusory or beneficial only to the coastal elites. The political challenge will be to sustain American competitive advantage and to distribute the benefits of U.S. prosperity more broadly. In this respect, three areas demand particular attention from policymakers: jobs and job retraining, aging, and innovation.

U.S. Vulnerabilities

Disruption, acceleration, and innovation are reshaping the global economy, creating opportunities for the United States, but also eroding the barriers to entry for new players. Calculations and transactions that once took days are now measured in gigabits per second. Major industries long ago developed global supply chains—what defines an American automobile these days? Suppliers compete in markets across

---

[22] Andrea Navarro, "Billionaire Immigrants Who Struck it Rich in the U.S.," Forbes, blog, March 19, 2014.

[23] According to the power cycle theory, pioneered by Charles Duran, a country's power can be defined as a share of all of the power of all of the countries in the world. In this analysis, a country can be "in decline" in relative terms but still be the world's leader with its power escalating in absolute terms. However, at inflection points in history, "everything suddenly and irremediably changes" and sudden discontinuity in foreign policy expectations can cause huge political uncertainty and increase the probability of major war. See Charles Duran, "Power Cycle Theory, the Shifting Tides of History, and Statecraft: Interpreting China's Rise," *SAIS Europe Journal of Global Affairs*, April 1, 2012.

the globe; competitive advantage is fleeting.[24] New technologies allow enterprises to decentralize as never before. Where humans once moved in search of jobs and opportunities, jobs now move in search of humans and opportunities, creating bubbles of prosperity at their destinations and busts in the areas left behind.

Since 1991, the global economy has swelled by 3.5 billion people, particularly from the nations of the former Soviet Union; from India, following the International Monetary Fund (IMF) reform programs it began that year;[25] and from China, whose economic integration accelerated after Chinese leader Deng Xiaoping's landmark southern tour in 1992 to restart economic reforms. The sheer size of the labor force in China and India changed the global labor equation in ways that went far beyond the influence of U.S. policymakers or the voters they serve. For example, India alone added 161 million people to its labor force between 1991 and 2014, and China added 241 million people. Together, this represents five times the increase in the entire U.S. population (68 million) during that same period.[26] In the United States, well paying, lower-skill jobs evaporated over this period, and not only in manufacturing.

The Great Recession compounded job insecurity. While the 8.7 million jobs that were lost during the recession had been regained by 2014, one study calculated that the recession also cost the United

---

[24] Rita Gunther McGrath and Alex Gourlay, *The End of Competitive Advantage: How to Keep Your Strategy Moving as Fast as Your Business*, Boston, Mass.: Harvard Business Publishing, June 4, 2013; Steve Denning, "It's Official! The End of Competitive Advantage," Forbes.com, June 2, 2013.

[25] Arvind Panagariya, *India's Economic Reforms: What Has Been Accomplished? What Remains to Be Done?* Manila, Philippines: Asian Development Bank, ERDC Policy Brief No. 2, November 2001.

[26] India's total labor force grew from 336 million in 1991 to 497 million in 2014, while China's grew from 645 million to 806 million, and the U.S. population increased from 251 million in 1991 to nearly 319 million in 2014 (World Bank, 2016a; U.S. Census Bureau, "Intercensal Estimates of the United States Resident Population by Age and Sex," Washington, D.C., 1991). Immigrants to the United States constituted 53.1 percent of U.S. population growth between 1990 and 2000, falling to 30.7 percent between 2010 and 2014, presumably due to the weak economy. This excludes additional population growth from immigrants' children. Shatz, 2016, pp. 49–50.

States economy 6.4 million "missing" jobs that were never created.[27] Although unemployment rates are now low, labor-force participation has continued to decline since 2000, even among men and women of prime working age.[28] Since 1991, earnings have been mostly flat. Real median earnings rose only 9 percent over 25 years,[29] not including benefits, such as employer-paid health insurance and other forms of compensation generally offered to better-paid workers. The percentage of Americans engaged in "alternative work arrangements"—temporary workers, on-call workers, freelancers or contractors—has also risen from 10.1 percent of the workforce in 2005 to 15.8 percent in 2015.[30] More than one in five adults were working more than one job, doing informal work as well as their main job, or both,[31] with some complaining about the lack of benefits and others welcoming the flexibility and earnings offered by this new "gig economy." Meanwhile, population mobility, long a driver of economic advancement in the United States, has decreased for reasons that are not yet understood, exacerbating the unevenness of economic opportunity.[32]

In contemporary U.S. politics, Chinese competition is often cited as the chief culprit in eliminating millions of well-paying factory jobs

[27] Anthony P. Carnevale, Tamara Jayasundera, and Artem Gulish, *Six Million Missing Jobs: The Lingering Pain of the Great Recession*, Washington, D.C.: Georgetown University Center on Education and the Workforce, 2015.

[28] There is considerable debate over the causes of the decline in labor force participation, some more worrisome than others. Factors include young people staying in school longer, young people unable to find jobs, more people retiring at a normal rate due to population aging, more demoralized older people retiring early and/or more people applying for disability payments. There is broad support for policies to improve the labor force participation rate as a means to boosting economic growth. See Ravi Balakrishnan et al., "Lost Workers," International Monetary Fund, *Finance and Development*, Vol. 52, No. 3, September 2015.

[29] Shatz, 2016, p. 20.

[30] Lawrence F. Katz and Alan B. Krueger, *The Rise and Nature of Alternative Work Arrangements in the United States, 1995—2015*, Harvard University, Princeton University and National Bureau of Economic Research, March 29, 2016.

[31] Consumer and Community Development Research Section of the Federal Reserve Board's Division of Consumer and Community Affairs, 2016.

[32] Raven Molloy et al., "Understanding Declining Fluidity in the U.S. Labor Market," Brookings Paper on Economic Activity Conference Draft, March 10–11, 2016.

that provided a middle-class lifestyle for Americans with a high school education. Certainly, low-cost imports from China and other countries have contributed to the loss of factory jobs, but even economists who see free trade agreements as a major factor in job loss estimate that imports were the cause of only about a quarter of all U.S. manufacturing job losses between 2000 and 2007,[33] or half, at most.[34]

Rather, increasing automation and other forms of advanced technology are deemed by many analysts to be far more significant than trade factors in the "creative destruction" occurring in the U.S. labor market and soon, even in lower-cost labor markets around the world.[35] As T. X. Hammes argues:

> The emergence of automated or "dark" factories that need only a few humans to supervise and maintain robotic production lines is a global trend. A fully automated factory in Mexico needs only six people per shift to produce thousands of cases of beer. The Changying Precision Technology Company in China has established an automated truck manufacturing plant that employees 90 percent fewer people. According to the Boston Consulting Group, about 10 percent of all manufacturing is currently automated, but this figure will rise to 25 percent by 2025.[36]

The "robots are killing millions of jobs" narrative in the U.S. media has produced decidedly mixed commentary about whether jobs

---

[33] See, for example, David H. Autor, David Dorn, and Gordon H. Hanson, "The China Syndrome: Local Labor Market Effects of Import Competition in the United States," *American Economic Review*, Vol. 103, No. 6, October 2013.

[34] Peter Gosselin and Mike Dorning, "After Doubts, Economists Find China Kills U.S. Factory Jobs," Bloomberg, June 18, 2015.

[35] For example, between 1999 and 2011, the newspaper, book, and directory publishing industry shed about 292,000 jobs while telecommunications shed about 397,000. Competition from China was not a factor in the loss of well-paying jobs in those industries, but technology was. U.S. Bureau of Labor Statistics, "Current Employment Statistics," database, series ID CES5051110001 and series ID CES5051700001, undated.

[36] T. X. Hammes, "Will Technological Convergence Reverse Globalization?" *Strategic Forum*, Institute for National Strategic Studies, National Defense University, SF No. 297, July 2016, p. 3.

menial enough to be performed by robots are worth saving at all,[37] and whether politicians have any realistic prospects of bringing them back to the United States. As the authors of *The Second Machine Age* argue, any routine task, whether manual or cognitive, is subject to replacement by automation.[38]

Some analysts argue that advances in manufacturing, including robotics, 3-D printing, and artificial intelligence, are already making some types of U.S. manufacturing cost-effective again and that this trend will sharply accelerate in the future.[39] New 3-D printing techniques, for example, can improve design, functionality, and even the performance of existing materials, while shortening supply chains and reducing shipping costs.[40] This trend toward reshoring is expected to create fewer jobs, though better-paid ones, for skilled workers who will design products, interpret data, repair automated machinery, and supervise the robots.

The evidence suggests that the technological advances that have greatly improved productivity—while creating and destroying jobs and industries—have also increased income inequality.[41] A recent IMF study, for example, notes that "technological changes can disproportionately raise the demand for capital and skilled labor over low-skilled and unskilled labor by eliminating many jobs through automation or

---

[37] For a flavor of how the debate is unfolding in the popular press, see Oscar Williams-Grut, "Robots Will Steal Your Job: How AI Could Increase Unemployment and Inequality," *Business Insider*, February 15, 2016; and Federico Pistono, "Robots Will Steal Your Job, But That's OK: How to Survive the Economic Collapse and Be Happy," *Business & Economics*, September 10, 2014.

[38] Brynjolfsson and McAfee further argue that advancing technologies increase "bounty"—the benefits associated with innovation—while also hastening the "spread" of inequality. Erik Brynjolfsson and Andrew McAfee, *The Second Machine Age*, New York: W.W. Norton, 2016.

[39] Hammes, 2016; Antoine van Agtmael and Fred Bakker, *The Smartest Places on Earth: Why Rustbelts Are the Emerging Hotspots of Global Innovation*, New York: Public Affairs, 2016.

[40] Hammes, 2016.

[41] For a discussion of the changing views of U.S. elites about income inequality, see Chris Matthews, "Even Harvard B-School Alums Are Fretting over Income Inequality," *Fortune*, September 9, 2015. For an international perspective, see the OECD work on inclusive growth, including *In It Together, Why Less Inequality Benefits All*, Paris: OECD Publishing, May 21, 2015.

upgrading the skill level required to attain or keep those jobs" and finds that advanced technology was responsible for nearly one-third of the growing income disparity in OECD countries over the past 25 years.[42] This suggests that absent policy changes, income inequality—within and between nations—will only increase.[43]

In the United States, however, despite the public focus on job dislocation and income inequality, the story is considerably more complex. For example, in December 2015, the Pew Research Center found that the American middle class was shrinking. What was less remarked upon was that this was because the shift from middle to upper was actually greater than the shift from middle to lower.[44] Upper-income households have gained the most, both in income (since 1970) and in wealth (since 1983)—so income inequality increased—but the share of adults living in upper-income households also increased.[45] Numerically, more Americans are doing better financially; yet this is not reflected in the anxiety expressed by many. Surveys conducted between 2014 and 2016 find high rates of financial stress even among those who were financially secure, and in spite of falling unemployment and an improving economy.[46]

For the bottom third of Americans, financial insecurity appeared to be a major factor. A 2016 Federal Reserve survey found that 46 percent of adults said they could not cover a $400 emergency expense,

---

42 Era Dabla-Norris et al., *Causes and Consequences of Income Inequality: A Global Perspective*, Washington, D.C.: International Monetary Fund, June 2015, pp. 18–20.

43 Brynjolfsson, 2016.

44 Pew Research Center, *The American Middle Class is Losing Ground: No Longer the Majority and Falling Behind Financially*, Washington, D.C., December 9, 2015b.

45 The study defined "middle income" as $42,000 to $126,000 in income for a three-person household in 2014.

46 Surveys indicated that financial anxiety is not confined to low-income Americans; millionaires also express a large degree of fear of losing their wealth and millennials who are millionaires are less likely to think they have "made it" and nearly twice as likely to be fearful of losing it than older ones. See "When is Enough . . . Enough? Why the Wealthy Cannot Get Off the Treadmill," UBS Investor Watch, 2Q 2015; Dave Shaw, "The Economy's Improving but Americans' Economic Anxiety Persists," *Marketplace*, March 14, 2016.

and 31 percent had no retirement savings at all.[47] Overall, 69 percent of adults said they were "living comfortably" or "doing okay" in 2015, up from 62 percent in 2013—and the improvement was reported even among households headed by someone with a high school education or less. At the same time, 31 percent said they were "struggling to get by" or "just getting by."[48] This failure to thrive among nearly one-third of the U.S. population raises the question of how to improve job prospects, economic well-being, and financial security without hurting the incentive systems that keep the U.S. economy strong.[49] Whatever future administrations' policies toward the "1 percent" of richest Americans, its policies toward the "31 percent" may prove most crucial in restoring political unity, as well as the U.S. position in the world.

Skills and Education

University and postgraduate education has been a traditional strength, and one that will be increasingly vital to U.S. international competitiveness. By one estimate,[50] the world faces a potential shortage of 38 million to 40 million college-educated workers as early as 2020, or a 13-percent undersupply, while it will have a potential surplus of 90 million to 95 million low-skill workers, roughly a 10-percent oversupply. This mismatch is expected to be particularly acute in the developing world, but employers in the advanced economies are projected to need 16 million to 18 million more college graduates than will be produced by 2020, despite rising college completion rates. The United States is projected to lack 5 million workers with postsecondary edu-

---

[47] Consumer and Community Development Research Section of the Federal Reserve Board's Division of Consumer and Community Affairs, 2016.

[48] Fifteen percent of households said they were spending more than they earned.

[49] Francis Fukuyama defined the challenge somewhat differently, as "to see whether it is possible to back away from globalization without cratering both the national and the global economy, with the goal of trading a little aggregate national income for greater domestic income equality." Francis Fukuyama, "American Political Decay or Renewal? The Meaning of the 2016 Election," *Foreign Affairs*, Vol. 95, No. 4, July/August 2016, p. 67.

[50] Richard Dobbs et al., *The World at Work: Jobs, Pay and Skills for 3.5 Billion People*, McKinsey Global Institute, June 2012, p. 48.

cation by 2020 (unless supply increases).[51] As seen in Table 4.1, trend lines indicate the United States, which is currently graduating only 2.8 million students with bachelor's, master's, or doctoral degrees per year,[52] will not increase the number of college graduates commensurate with expanding demand.

These figures do not take into account the smaller number of students graduating in science, technology, engineering and mathematics (STEM), the skills expected to be most in demand from employers. Only about 28 percent of college undergraduates enter STEM fields, and attrition from those academic programs is high. Even though starting salaries for STEM graduates are much higher than for other fields, employers say job openings in STEM fields are harder to fill.

Moreover, at current rates, immigration will not be sufficient to plug the skills gap in the advanced economies.[53] The U.S. economy may continue to benefit, as it has in the past, from its ability to employ immigrants, but it will clearly need to produce more highly educated people, both to fill jobs and to create higher-paying ones. Even as

Table 4.1
Expanded Supply of U.S. Graduates in Higher Education

Years	Bachelor's Degrees	Master's Degrees	Doctoral Degrees
2015–2016	1,846,000	802,000	179,000
2020–2021	1,933,000	920,000	198,000
2024–2025	2,029,000	1,019,000	209,000

SOURCE: National Center for Education Statistics, "Digest of Education Statistics," Table 318.10, "Degrees Conferred by Postsecondary Institutions, by Level of Degree and Sex of Student: Selected Years, 1869–70 through 2024–25," web page, April 2015.

---

[51] Anthony P. Carnevale, Nicole Smith, and Jeff Strohl, *Recovery: Job Growth and Education Requirements Through 2020*, Washington, D.C.: Georgetown Public Policy Institute, Georgetown University, June 2013, pp. 15–21. The United States has been producing too few highly educated workers since the 1980s. The authors estimate that 65 percent of all jobs will require postsecondary education by 2020, up from 28 percent in 1973; 26 percent will require less than a high school degree or a high school diploma; 30 percent will require some college or an associate's degree; 24 percent will require a bachelor's degree, and 11 percent will require a master's degree or more.

[52] National Center for Education Statistics, 2015.

[53] Dobbs et al., 2012, p. 46.

employment improves, structural problems in the U.S. labor force persist. As of April 2016, the United States had a notable gap between the number of job openings (on average, more than 5 million) and the number of people who are unemployed (about 8 million, of whom more than a quarter have been unemployed for six months or more).[54] This does not include the number of people working part-time involuntarily (6 million, as of April 2016) or those who have are dropped out of the labor force entirely (568,000 "discouraged workers" who have stopped looking for jobs) or the 1.7 million more people who had looked for work in the past year but not in the past month, and so were not counted in the official unemployment rate.

There is vigorous debate over the extent to which the so-called "skills mismatch" argument explains this phenomenon of unfilled jobs even during periods of high unemployment.[55] How many Americans are not working because they lack the skills that employers demand? Other possible causes include wages that are too low to attract available workers (hourly wages have not rebounded even though employment has), a reluctance by employers to invest in job training programs, family responsibilities, ill health, lack of transportation,[56] unwillingness to relocate, inability to pass a drug screening test,[57] or an arrest or criminal record.[58]

---

[54] U.S. Bureau of Labor Statistics, 2016.

[55] See, for example, Gary Burtless, "Unemployment and the 'Skills Mismatch' Story: Overblown and Unpersuasive," Brookings Institution, blog, July 29, 2014; and Elise Gould, "Still No Sign of a Skills Mismatch—Unemployment is Elevated Across the Board," Economic Policy Institute, blog, January 13, 2015; and Tyler Cowan, "Skill Mismatch Unemployment is Real and Significant," MarginalRevolution, website, November 21, 2015.

[56] Dobbs et al., 2012.

[57] Jackie Calmes, "Hiring Hurdle: Finding Workers Who Can Pass a Drug Test," *New York Times*, May 17, 2016.

[58] The Federal Bureau of Investigation had more than 77 million individuals on file in its master criminal database in 2014. Gary Fields and John R. Emshwiller, "As Arrest Records Rise, Americans Find Consequences Can Last a Lifetime: Even When Charges Are Dropped, a Lingering Arrest Record Can Ruin Chances of a Job," *Wall Street Journal*, August 18, 2014.

For those who are working, however, the wage premium for college graduates has soared, particularly since 2000.[59] Between 2000 and 2009, a person with a bachelor's degree earned 127 percent of what a person without a high school diploma did, and a person with an advanced degree made 187 percent more. This wage gap became even steeper after the Great Recession.[60] The trend toward income inequality was accentuated by the notable growth in so-called "superstar earners." This refers to the top 0.1 percent of earners in the United States, whose share in the population quadrupled between 1976 and 2012, from 2 percent to 8 percent.[61] This trend, which is also seen in other advanced economies, is partly explained by technology and globalization, but other factors are also believed to be at play. In the United States, a range of extended poverty programs have cushioned the effects of inequality; while the official poverty rate has increased since 2000, tax credits and transfer payments kept the real poverty rate from rising.[62] The increase in child poverty rates has become a major concern, as the rate for children under six years of age, which had fallen to 17.8 percent in 2000, was 23.5 percent in 2014.[63] Parental income is strongly correlated with the scholastic achievement of children.[64] How-

---

[59] The wage premium—the extra amount that college-educated workers earned compared with non–college-educated workers, after controlling for experience and gender—averaged around 50 percent between 1963 and the early 1980s but then rose sharply through 2012 to peak at nearly 100 percent. Charles I. Jones, *The Facts of Economic Growth*, Cambridge, Mass.: National Bureau of Economic Research, NBER Working Paper No. 21142, 2015, p. 17.

[60] Shatz, 2016, pp. 18–19.

[61] Francisco Perez-Arce et al., *Inequality and Opportunity: The Relationship Between Income Inequality and Intergenerational Transmission of Income*, Santa Monica, Calif.: RAND Corporation, RR-1509-RC, 2016, pp. 54–55.

[62] Perez-Arce et al., 2016, pp. 19–20.

[63] U.S. Census Bureau, "Historical Poverty Tables: People and Families—1959 to 2015," web page, Table 20, "Poverty Status of Related Children Under 6 Years of Age," September 2016a.

[64] The gap between the test scores of the poorest and richest 10 percent of the U.S. population among students born in 2000 was 75 percent wider than among students born in 1943. Edward Alden and Rebecca Strauss, *How America Stacks Up: Economic Competitiveness and U.S. Foreign Policy*, New York: Council on Foreign Relations, 2016.

ever, there are no direct measurements to gauge the current impact of policies aimed at combating inequality of opportunity.[65]

Compared with the rest of the world, the United States has a lower preschool enrollment rate and higher college drop-out rate—and, in contrast to the previous generation, the college graduation rate for Americans ages 25 to 34 ranks 12th in the world.[66] Meanwhile, enrollment in apprenticeship programs to prepare young people for jobs has declined 40 percent in the last decade, although these have proved successful in Germany and other nations.[67] While there is strong U.S. public support for job retraining and education for the unemployed, the data are scanty on which retraining programs are effective and which are not. In one study of 47 programs between 2004 and 2011, a comprehensive impact study was conducted in only five.[68]

Raising education levels will not guarantee better living standards for all Americans. Failure to educate and retrain, however, will mean that fewer Americans will have the skills to succeed in an advanced economy facing rising levels of global competition.

Recruiting and Retaining Global Talent

U.S. universities are a beacon for international students, although they are unaffordable for many and many graduates who wish to stay cannot do so.[69] The U.S. knowledge economy will benefit most from attracting and training top talent that will stay in the United States. Proposals for policies to accomplish this include increasing the number of H-1B non-immigrant visas for high-skilled technical workers who

[65] Perez-Arce et al., 2016.

[66] Alden and Strauss, 2016.

[67] Alden and Strauss, 2016.

[68] Alden and Strauss, 2016.

[69] One well-publicized case was Kunal Bahl, who graduated from Wharton School but was deported in 2008. He went back to India and founded Snapdeal.com, a technology company now valued at more than $5 billion. Dina Bass, "America's Unwanted Ivy Leaguers Are Flocking to India," Bloomberg, June 2, 2015.

will remain in the United States;[70] training and supporting talented students in STEM programs who will also remain in the United States for a defined period; and offering "start-up visas" for immigrants who want to launch a business in the United States, as about 15 other countries have done.[71] The payoff appears to be rapid: While immigrants make up less than 13 percent of the U.S. population, between 1995 and 2005, more than 25 percent of all new engineering and technology companies had at least one foreign-born co-founder.[72] Yet the number of H-1B visas issued has been reduced.

An increase in H-1B visas might be paired with increasing funding for higher education of American students in STEM fields. The next administration might consider pilot programs designed to test the effectiveness of these and other proposals for improving the skills of the U.S. workforce.

Aging

The U.S. workforce is aging better than its friends or rivals. It is the only one of the large, affluent nations that is projected to see increases, however modest, in its working age population,[73] due to a combination of fertility rates and immigration. Its demographic curve skews younger than that of Europe, Japan, South Korea, China, or Russia but not younger than that of many of the least-stable Islamic countries—including Afghanistan, Yemen, Egypt, Nigeria, Iraq, Sudan, and Pakistan—where the large percentage of young males (ages 15 to 25) in the population, along with poor employment prospects, are considered by some to pose a threat to stability. (Birthrates in many sub-Saharan African nations have not declined as quickly as had been expected, generating fears that these

---

70 David Petraeus and Paras D. Bhayani, *The Next Great Emerging Market? Capitalizing on North America's Four Interlocking Revolutions*, Cambridge, Mass.: Belfer Center, Harvard Kennedy School, June 2015.

71 Fairlie et al., 2015.

72 Brynjolfsson, 2016, pp. 222–224.

73 Martin C. Libicki, Howard J. Shatz, and Julie E. Taylor, *Global Demographic Change and Its Implications for Military Power*, Santa Monica, Calif.: RAND Corporation, MG-1091-AF, 2011.

countries may not be able to raise living standards as rapidly as had been hoped and thus might face continuing instability).

In contrast, because of its birth and immigration rates, the American working-age population is projected to remain in the "Goldilocks" range, edging down from 4.7 percent of the world's working-age population in 2011 to 4.3 percent in 2050.[74]

Demography does not destine the United States to economic decline. A 2011 RAND study concluded that "there is no reason to believe, at this point, that population aging is likely to flatten economic growth rates." It is certainly true that the retiring Baby Boomers will strain the federal budget through increasing demands on Social Security and Medicare. However, this depends on how many choose (or are forced) to retire early, and this, in turn, will be influenced by the political choices to be made about the age of eligibility for pensions or benefits or other incentives for work. The sooner such policy decisions are made, the sooner employers and workers can begin planning to adapt.

Labor productivity is another wild card. Will older workers stay on the job longer if robots are doing most of the heavy lifting?[75] Can older workers be retrained to run them?[76] Automation leading to increases in productivity and retention of highly skilled workers in some industries could boost U.S. growth, offsetting the job-killing effects of other types of automation and technological advance.

Medical costs for an aging population are expected to rise. Even so, choices made by individual Americans and policymakers can affect these outcomes greatly. As Martin C. Libicki put it:

---

[74] The overall percentage of the working-age population in Muslim-majority countries is projected to rise to 28.2 percent in 2030 and 30.7 percent in 2050.

[75] Robotic exoskeletons—motorized suits that lend the wearer strength, stamina, or protection—are being developed for use in industry, the military, and by disabled people. For examples, see "Overview of Robotic Exoskeleton Suits for Limb Movement Assist," Smashing Robotics, May 6, 2016.

[76] "If older workers are trainable, are inventive, and can absorb new technology and methods, labor productivity trends could continue along the same path as before. If they are not, economic growth may suffer." Libicki, Shatz, and Taylor, 2011, p. 57.

> Notwithstanding any other factor, the changing age distribution of the U.S. population would raise per capita health expenditures by 20 percent between 2000 and 2030. But one cannot ignore these "other factors."[77]

These include disease patterns, entitlement to medical treatment, economic shifts, technological changes, and social changes, including changes in diet, lifestyle, smoking, suicide rates, and opioid and alcohol use. These factors are within the country's power to change to some degree, but as Case and Deaton warned:

> A serious concern is that those currently in midlife will age into Medicare in worse health than the currently elderly. This is not automatic; if the epidemic [of increased morbidity and mortality due to opioids, alcohol, and suicide] is brought under control, its survivors may have a healthy old age.[78]

A particularly negative factor in the aging U.S. population is the toll taken by dementia, a disease that now costs the country more than heart disease or cancer. A RAND study has estimated the monetary cost of dementia at $157 billion to $215 billion per year.[79] Assuming the prevalence of the disease remains steady, with roughly three in 20 Americans ages 71 or older affected, these costs may more than double by 2040, due to the increased number of older Americans. However, there is some evidence to suggest that the age-specific dementia rate is declining, in the United States and four other industrial countries, possibly due to rising levels of education and more-aggressive treatment of cardiovascular conditions that are risk factors for the disease.[80] However, high rates of obesity and diabetes could counteract this positive trend.

---

[77] Libicki, Shatz, and Taylor, 2011, p. 59.

[78] Case and Deaton, 2015, p. 4.

[79] Michael D. Hurd et al., "Monetary Costs of Dementia in the United States," *New England Journal of Medicine*, Vol. 368, No. 14, April 4, 2013.

[80] Kenneth M. Langa, "Is the Risk of Alzheimer's Disease and Dementia Declining?" *Alzheimer's Research and Therapy*, Vol. 7, No. 1, March 26, 2015.

As with proposals to reform education, there are a large number of recommendations for early interventions in public health that may protect the health of aging adults, thereby lessening the cost of preventable illness and disability in an aging population. However, absent changes to the cost and efficacy of American medicine, or robust evidence-based public health interventions to improve health outcomes, U.S. health care costs will consume an ever-larger share of GDP as the population ages. Mandatory spending on such programs as Medicare, Medicaid, and Social Security consumed 12.9 percent of GDP in 2015.[81] Discretionary spending continues to shrink, falling from 9 percent of GDP in 1991 to 6.5 percent of GDP in 2015.[82] At some point, lack of discretionary funds becomes a constraint on the exercise of U.S. military and nonmilitary power abroad.

Innovation

Innovation remains the most critical source of U.S. economic strength and competitive advantage, just as creativity remains a powerful component of the value of the American "brand." Other countries are catching up fast. They are challenging U.S. dominance in STEM skills, as well as American cultural leadership. As part of national strategy, the United States needs to provide a climate that can nurture innovation, entrepreneurship, and productivity gains.

Much of this depends on the private sector. However, as shown in Figure 4.1, U.S. federal investment in research and development (R&D) as a percentage of GDP has been declining over the past three decades, although absolute spending has increased. Significantly, both military and civilian R&D declined during a period where other countries began to increase their spending.

---

[81] Maureen Costantino and Leigh Angres, "The Federal Budget in 2015," fact sheet, Congressional Budget Office, Washington, D.C., January 2016.

[82] Jonathan Schwabish and Courtney Griffith, "The U.S. Federal Budget: A Closer Look at Discretionary Spending," fact sheet, Congressional Budget Office, Washington, D.C., April 2012.

Figure 4.1
Decline in Federal R&D Spending

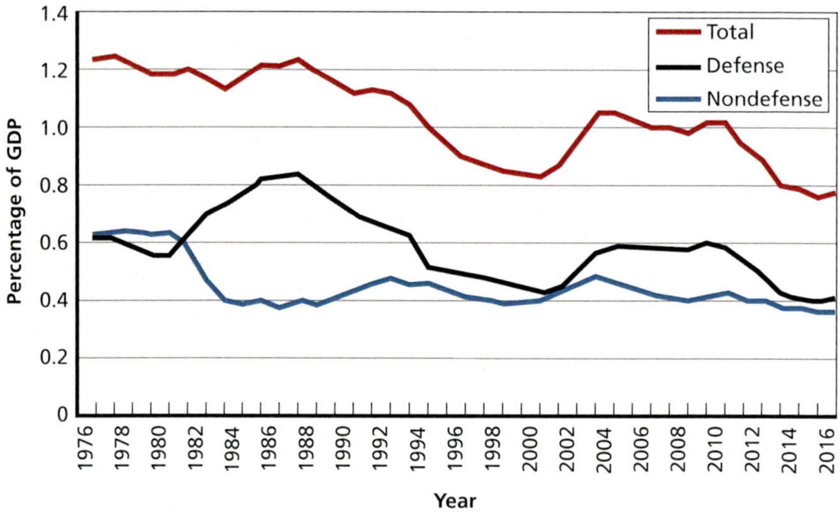

SOURCES: Up to 1994: National Science Foundation, *Federal Funds for Research and Development Detailed Historical Tables: Fiscal Years 1956–1994*, Bethesda, Md., NSF 94-331, 1994. 1995 to present: American Association for the Advancement of Science, "Historical Trends in Federal R&D," web page, 2015. GDP figures: Office for the Under Secretary of Defense (Comptroller), *National Defense Budget for FY 17 (Green Book)*, Washington, D.C.: U.S. Department of Defense, March 2016.
NOTE: 2015 and 2016 figures are estimates.
RAND *RR1631-4.1*

The United States remains the largest spender on R&D, investing 2.8 percent of GDP in 2014.[83] However, the U.S. share of global R&D spending has declined slightly, while China's share has risen, resulting in estimates that China's R&D spending will overtake the United States' by 2022.[84]

Significantly, the percentage of global R&D conducted in Asia (35 percent, primarily in China and India) now slightly exceeds that

---

83 "2014 Global R&D Funding Forecast," *R&D Magazine*, December 9, 2013, p. 4.

84 "2014 Global R&D Funding Forecast," 2013, p. 3. The National Science Foundation found that the U.S. share of global R&D declined from 37 percent in 2001 to 30 percent in 2011. National Science Board, "Science and Engineering Indicators 2014," National Science Foundation, Arlington, Va., NSB 14-01, February 2014, Chapter 4.

conducted inside the United States (33 percent), a trend that is increasing as companies headquartered in the United States and Europe move research operations offshore in pursuit of lower costs and closer markets.[85] In 2015, Volkswagen and Samsung led the world in corporate R&D, spending significantly more than Microsoft, Google, Amazon, Toyota, or Apple.[86]

In general, private-sector investments tend to pay off more quickly because they aim to generate goods and services that are salable or likely to be so in future. In this sense, "salability" is an immediate indicator that a given technology is of actual value. Tax and other federal policies that increase incentives for private-sector R&D in the United States may be desirable to the extent that they spur the creation of valuable intellectual property. However, for-profit firms do not invest in basic research that does not have the prospects for yielding salable products in a relatively short span of time. Strategic federal R&D programs attempt not to "pick winners and losers" among specific technologies, but rather to support basic scientific research in areas where evidence suggests that commercial firms will not invest, either because the likely profit margins are too low or because the time line for return on investment is too long.

Well-conceived policies to promote U.S. productivity would be desirable, if not essential, under at least two of the three strategies outlined in this report (Options II and III). Further, important technological developments or scalability may not be achieved on the time line desirable to the U.S. government in light of the particular, rapidly evolving strategic challenges already discussed. Therefore, if pursuing these strategies, the United States may wish to be more activist in spurring basic research that can underwrite productivity gains and innovation in areas where the private sector is unlikely to invest sufficient R&D funds at sufficient speed and scale.

American private enterprise has shown itself to be innovative and efficient; the public sector much less so. The federal government has

---

85 Barry Jaruzelski, Kevin Schwartz, and Volker Staack, "The 2015 Global Innovation 1000: Innovation's New World Order," *Strategy and Business*, No. 81, Winter, October 27, 2015.

86 Jaruzelski, Schwartz, and Staack, 2015, Exhibits 8 and 9.

struggled to adapt private-sector technologies to improve performance, and DoD would likely benefit from using more private-sector technology and adapting best commercial practices. DoD defense and space programs have satisfied U.S. defense needs and generated valuable spin-offs, however inefficient the process of creating defense-related technology may be. Former Federal Reserve Chairman Ben Bernanke recently remarked that the U.S. defense budget is what passes for industrial policy in the United States.[87] And these defense investments have yielded tremendous civilian rewards, including the creation of the Internet and the driverless car, both born as a result of funding at DoD's Advance Research Projects Agency, and the Global Positioning System, developed by the military. Defense Secretary Ashton Carter has made a special effort of reaching out to Silicon Valley. Regardless of its views on economic and industrial policy, the next administration should consider investing more in basic scientific research across the defense and civil sectors.

The Obama administration's emphasis on stimulating innovation in carbon-free energy technology is expected to be slowed or reversed by the Trump administration, which has promised to increase domestic energy exploitation.[88] The revolution in extraction technologies, commonly referred to as "fracking," increases global supplies and therefore helps reduce the price of petroleum, increases availability and lowers the cost of natural gas as an input to manufacturing, and reduces the U.S. trade deficit. But even as its energy imports decline, the United States will remain exposed to global supply and price volatility because the world market for petroleum is unified.

However, any breakthrough on zero-carbon energy technologies would be a game-changer. Private and philanthropic capital is already flowing to such projects. The strategic and political question for the next administration is whether it is essential to the United States that the leading forms of green-energy technologies be developed and patented by American citizens and/or developed by U.S. corporations.

---

[87] Ben S. Bernanke, "Proceedings," The Defense Economy and American Prosperity, Washington, D.C., August 17, 2015.

[88] Trump, 2016.

If it considers such technologies to be crucial to U.S. competitiveness, the next administration might consider supporting a carbon tax or cap-and-trade market system to incentivize low-carbon technology development and to promote basic scientific research and advancement of knowledge for all, as opposed to providing federal funding for research on specific energy technologies (an approach often criticized as a futile attempt to "pick winners"). If not, the United States must accept the risk that its competitors, including European and Asian nations that have adopted "green" industrial policies, may pioneer breakthrough technologies.

Following this line of reasoning, the next administration also might expand research on basic science relevant to medicine and health policy, focusing on R&D of treatments and devices that advance public health but are unlikely to generate short-term profits.[89] This may be beneficial for a number of reasons. Domestically, U.S. health care costs will consume an ever-larger share of GDP as the population ages, absent changes to the cost and efficacy of American medicine. Internationally, medical knowledge and technology present both a commercial growth market and a realm where U.S. interventions have alleviated misery and generated global goodwill.

Public Opinion

The overarching question of how deleterious the effects of income inequality and job displacement are on America's ability to lead globally is beyond the scope of this discussion. The immediate problem for national strategy is whether the United States can maintain a robust, far-sighted foreign policy if a significant portion of the electorate believes that the economic components of that policy are not advanc-

---

[89] The Gates Foundation funding for vaccine development is one example. Another comes from the not-for-profit Benetech, which adapts for use in humanitarian endeavors technologies that were developed by the private sector but abandoned as not economically promising. Various forms of government partnerships with so-called "social entrepreneurs" matching grants for private investment in other start-ups and similar attempts to use government financing as a bridge to private venture capital are under way through the U.S. Small Business Association.

ing the interests of the majority.[90] Given a choice between domestic spending and defense and foreign assistance, public opinion data show a strong preference for cutting defense spending and spending at home on issues that are seen as directly affecting well-being, such as education, jobs, training, and energy. Jennifer Kavanagh notes that popular and elite attitudes frequently diverge on such matters, with popular opinion closely correlated with personal experience and observation.[91] For example, supporters of foreign trade liberalization are those in professional jobs who are helped by low prices and those who oppose it are blue-collar workers who lose job security.

Simply put, voters make decisions on personal experience and concrete evidence, and their attitudes on abstract or peripheral issues are weaker and more ambivalent. Therefore, Kavanagh argues, public opinion is a "sleeping giant that is docile on some issues but difficult to rein in once awakened and potentially dangerous when activated. Where it crystallizes, public attitudes appear to be significantly influenced by personal interests."[92]

Historically, a number of U.S. presidents have maneuvered the country into supporting policies that were highly unpopular at the time, risking awakening the "sleeping giant" of a public unpersuaded that the President's preferred policy would advance their personal inter-

---

[90] Two of the most heated political issues are trade and government help for the middle class. In a survey of 1,500 respondents conducted December 8–13, 2015, by the Pew Research Center, 62 percent said the federal government does not do enough for middle-class Americans, while 6 percent said it does too much. Pew Research Center, *Most Americans Say Government Doesn't Do Enough to Help Middle Class*, Washington, D.C., February 4, 2016a. Exit polling in early 2016 found that a majority of Republican primary voters and a plurality of Democratic voters surveyed believed that trade takes away U.S. jobs. Gregory Holyk, "Foreign Policy in the 2016 Presidential Primaries Based on the Exit Polls," Chicago Council on Global Affairs, April 7, 2016.

Yet the Gallup organization's tracking polls found that support for trade, which has fluctuated greatly in recent decades, has been rising since the Great Recession—as of February 2016, 58 percent saw trade as an opportunity, compared with 34 percent who saw it as a threat. Justin McCarthy, "Americans Remain Upbeat About Foreign Trade," Gallup, February 26, 2016.

[91] Interview with RAND political scientist Jennifer Kavanagh, June 6, 2016.

[92] Interview with Kavanagh, 2016.

ests. Most famously, Franklin D. Roosevelt and his Secretary of State Cordell Hull pushed through the Lend-Lease program to help Britain in January 1941, 11 months before the Japanese attack on Pearl Harbor. They quietly prepared the country for the Second World War at a time when public opinion strongly opposed U.S. entry into the European conflict. In 1964, Johnson seized on the alleged North Vietnamese attacks on U.S. naval forces in the Tonkin Gulf to win congressional mandate for expanding the Vietnam War, which was already unpopular.[93] Johnson and Nixon staunchly continued the containment policies of their predecessors despite mounting opposition during the Vietnam War. George W. Bush commanded a troop "surge" into Iraq in 2007 to stop sectarian bloodletting and stabilize the Iraqi government even as domestic support for the war plummeted. However, the United States has been strongest, as it was during the early years of the Cold War, when it has been able to sustain a considerable measure of consensus for a broader national purpose among government, business and labor, and social classes. Such broad consensus among the sources of power in the emerging new economy has been notably absent in this century, except for a brief period following the 9/11 attacks.

The collapse of consensus over trade policy has been particularly sudden. Labor and the political left had argued for 40 years that unfair trade was destroying American jobs. The 1970s saw widespread domestic opposition to imported Japanese automobiles. The 1980s brought sharp trade friction with Japan. The 1990s brought accusations that Japan, Brazil, and other nations were destroying the U.S. steel industry by dumping steel on the American market.[94] The 2000s brought warnings

---

[93] The Johnson administration was later accused of having distorted the facts about the North Vietnamese attacks. The attack on the U.S. destroyer *Maddox* on August 2, 1964, was confirmed but there was confusion about the alleged August 4 attack on the *Turner Joy*. Defense Secretary Robert McNamara later concluded that no attack had occurred. Lt. Comm. Pat Paterson, "The Truth About Tonkin," *Naval History Magazine*, U.S. Naval Institute, Annapolis, Md., Vol. 22, No. 1, February 2008; Errol Morris and Robert S. McNamara, interview, *The Fog of War*, Sony Pictures Home Entertainment, 2005.

[94] The United States has had anti-dumping statutes on the books since 1921 and the Commerce Department continues to impose tariffs when it deems there have been transgressions, most recently against Chinese solar panel manufacturers.

that U.S. manufacturing jobs were being exported to China.[95] In 2016, U.S. officials were once again pushing China to stop its alleged dumping of steel.[96] Nevertheless, bipartisan political support for free trade policies held firm through the Clinton, Bush, and Obama administrations, all of whom have cited compelling evidence that such trade deals benefited the United States in the aggregate. Following the slow recovery from the Great Recession, rising concern over U.S. deindustrialization,[97] wage stagnation, income inequality—and, some argue, trade-shocks that have particularly hurt low-wage workers[98]—culminated in 2016 with the presidential candidates of both major political parties for the first time declaring their opposition to President Obama's signature trade deal, the Trans Pacific Partnership (TPP).

The trade pact is now considered dead, and it is unclear whether President-elect Trump will succeed in renegotiating or repudiating NAFTA and other pro-globalization policies that have been favored by U.S. political and economic elites for seven decades. The answer may ultimately depend on what fraction of the U.S. workforce is permanently dislocated by the changing economy, and in what numbers they vote. What percentage of citizens will experience permanently lower standards of living because they have lost a good-paying job—or never managed to land one in the first place—as a result of poor education, foreign competition, technological advances, geographical location, or some combination of these factors? What percentage will exit the labor

---

[95] At the same time, manufacturing jobs also declined in China between 1994 and 2005. William H. Overhold, "China and Globalization," testimony to the U.S. China Economic and Security Review Commission, May 19, 2005.

[96] David Lawder, "Lew Says Excess Capacity 'Corrosive' for China Growth," Reuters, June 5, 2016.

[97] U.S. manufacturing employment declined 18 percent from March 2001 to March 2007. Justin R. Pierce and Peter K. Schott, *The Surprisingly Swift Decline of U.S. Manufacturing Employment*, Cambridge, Mass.: National Bureau of Economic Research, Working Paper No. 18655, December 2012, revised January 2014.

[98] This argument was made by David H. Autor, David Dorn, Gordon Hanson, and Jae Song, "Trade Adjustment: Worker-Level Evidence," *Quarterly Journal of Economics*, Vol. 129, No. 4, 2014.

market,[99] accept disability insurance, some form of trade adjustment assistance, or other antipoverty benefits as a substitute, and then "age out" of the workforce only to subsist on Social Security and Medicare?

Could future administrations design programs to cushion those who have not benefited from globalization in order to sustain free-trade arrangements that have boosted the gross domestic product, however unevenly those gains are distributed? Can this be done without massively increasing the federal deficit or the debt-to-GDP ratio, which also pose long-term challenges to U.S. competitiveness in the world?

Whatever the specific policies chosen, maintaining a climate to attract and promote entrepreneurship and investment, and building an economy that delivers on the promise of the American dream for a larger share of the population must be part of an "opportunity agenda" for any administration. Success at home begets strength abroad; it strengthens any U.S. president's power of persuasion. International strategy will be most successful when it reflects the recognition that the U.S. power to attract and inspire is greater than its power to compel.

---

[99] Long-term unemployment is correlated with a lower likelihood of subsequent employment. In 2014, about 35 percent of people who had been out of work for a year or more became unemployed in the following month. Karen Kosanovich and Elini T. Sherman, *Trends in Long-Term Unemployment,* Washington, D.C.: Bureau of Labor Statistics, March 2015.

CHAPTER FIVE

Anticipation, Deterrence, and Resilience

> The best way to keep something bad from happening is to see it ahead of time . . . and you can't see it if you refuse to face the possibility.
>
> William S. Burroughs[1]

The renowned physicist Nils Bohr, who developed a model to predict the seemingly random movement of electrons, liked to observe, "Prediction is very difficult, especially if it's about the future."[2] The inadequacy of prediction is like a law of physics, a constant in human affairs. The United States spends nearly $70 billion per year on intelligence,[3] of which at least $20 billion is for the purpose of warning civilian and military leaders about possible threats.[4] And yet, U.S. policymakers have been no

---

[1] Burroughs was a 20th century writer, philosopher, and an original member of the Beat Generation. "William S. Burroughs, Quotable Quote" goodreads.com, undated.

[2] Bohr, one of the founders of modern physics, attributed this quotation to Robert Storm Petersen (1882–1949), but its provenance is disputed. "Letters to the Editor: The Perils of Prediction, June 2nd," *The Economist*, July 15, 2007.

[3] In fiscal year 2015, the administration requested $50.3 billion for the National Intelligence Program and $16.8 billion for DoD's Military Intelligence Program. Federation of American Scientists, "Intelligence Budget Data," web page, 2016; Michael O'Hanlon, *U.S. Defense Strategy and the Defense Budget*, Washington, D.C.: Brookings Institution, November 2015, p. 4.

[4] Barton Gellman and Greg Miller, "'Black Budget' Summary Details U.S. Spy Network's Successes, Failures and Objectives," *Washington Post*, August 29, 2013; related documents at "Inside the 2013 U.S. Intelligence 'Black Budget,'" *Washington Post*, August 29, 2013.

less surprised by recent events—from 9/11 and the Arab Spring to the rise of ISIS and Russia's entry into the Syrian conflict—than their predecessors were 30 years ago when the United States failed to anticipate the collapse of the Soviet Union. Today's policymakers find themselves increasingly frustrated by jobs that involve more crisis management—suppressing system shocks—than strategic thinking,[5] and by the difficulty (especially in the current political environment) of mustering the consensus, trust, and thus the resources necessary to prevent and/or prepare even for those developments that can be foreseen.[6] The large influx of resources since 9/11 has helped the intelligence community in one of its primary missions: intelligence relevant to counterterrorism. However, it does not appear to have improved the U.S. ability to anticipate other kinds of developments or prevent surprise.[7] And many things that surprise us are not actually "intelligence failures"—rather, they result from a broader inability to pair foresight with action. Yet the more turbulent the times, the more acute the need for U.S. leaders to look ahead, contemplate actions that could be taken in the present to improve future outcomes, and anticipate events rather than respond to them.[8]

Contrary to popular belief and the needs of policymakers, intelligence is not about predicting the future. It is about collecting and analyzing information that helps explain the physical and political

---

[5] Julianne Smith, "Our Overworked Security Bureaucracy," *Democracy Journal*, No. 40, Spring 2016. She concludes that, "While it is no doubt impossible to predict how and when the next crisis will unfold, our government must do a better job of assessing risk, testing core assumptions, and preparing itself for potential contingencies."

[6] The U.S. difficulty in funding a response to the Zika virus is a recent example. Nora Kelly, "The Senate Goes Home Without Funding Zika," *The Atlantic*, April 29, 2016.

[7] Known strategies used to reduce surprise include relying on experience, reducing the number of variables in a problem, adopting a measured response to preserve future options, and teamwork. Dave Baiocchi and D. Steven Fox, *Surprise! From CEOs to Navy Seals: How a Select Group of Professionals Prepare for and Respond to the Unexpected*, Santa Monica, Calif.: RAND Corporation, RR-341-NRO, 2013.

[8] The Project on Forward Engagement has been working on a systems-based approach to "anticipatory governance" since 2011. Leon Fuerth and Evan M. H. Farber, *Anticipatory Governance, Practical Upgrades: Equipping the Executive Branch to Cope with Increasing Speed and Complexity of Major Challenges*, Washington, D.C.: The Project on Forward Engagement, October 2012.

world, and by doing so, helping leaders make better decisions. As Richard Betts wrote, "It is the role of intelligence to extract certainty from uncertainty and to facilitate coherent decision making in an incoherent environment."[9] As discussed in Chapter Three, globalization, the increasing numbers of world actors and variables, and the "Rubik's Cube" quality of the complex and concurrent interactions among them have increased *uncertainty*, defined as a state in which "information is too imprecise to be summarized by probabilities."[10] Uncertainty is an intractable problem for a nation that spends more than $600 billion per year to defend against security threats, actual and anticipated.

How can the U.S. government improve its ability to make better decisions in an uncertain and perhaps incoherent environment? We offer two suggestions. First, it should streamline the unwieldy bureaucratic apparatus for national security decisionmaking. This means a leaner National Security Council staff and planning apparatus that can extract the right conclusions from the deluge of analysis and intelligence and can carve out more time for thinking, prioritizing, and strategizing. A more agile decisionmaking structure is more important than ever as policymakers find themselves "drinking from the firehose:" trying to absorb vastly more information from many additional sources at a faster pace than ever before. Even though the world has changed, many of our governing structures have not. The national security decisionmaking apparatus is overstaffed and underperforms in some organizational matters, particularly compared with its leaner rivals. Success depends not just on wise strategic and policy choices, but also on mechanisms that will foster more agile and effective governance.[11]

Second, we suggest a deeper focus on three policymaking approaches that are particularly relevant in highly uncertain environ-

---

[9] Richard K. Betts, *Enemies of Intelligence: Knowledge and Power in American National Security*, Columbia, N.Y.: Columbia University Press, 2007, p. 30.

[10] Goldin and Mariathasan, 2014, p. 25. In the early 20th century, economist John Maynard Keynes and statistician Frank Knight were among the first to formally distinguish between *risk*, which can be quantified, and *uncertainty*, which cannot. In more recent years, the term *deep uncertainty* has been used for information that is too imprecise to be summarized with confidence or expressed by quantified probabilities.

[11] Ries, 2016.

ments, and that our analysis suggests can be improved: anticipation, deterrence, and resilience. We must anticipate what we can, and act on this foresight; rethink how to deter those threats we can anticipate; and build resilience to withstand and rebound from any attacks, surprises, or calamities that we cannot anticipate or deter.

Anticipation, deterrence, and resilience are common terms, but they are commonly misunderstood. *Anticipation* is foresight plus appropriate action. It is use of foresight to inform actions that can be taken now to shape the future. In other words, we act now to make a future scenario we do not want less likely, or to increase the odds of outcomes we do want. *Deterrence* also aims to shape the future: Based on the foresight that someone may do something that runs counter to our interests, we can take action now to make such actions more costly and thus less likely. *Resilience* is a hedging action: We know we cannot predict the precise forms of mayhem that may come our way, so we insure ourselves against unknown risks by preparing to withstand whatever may come.

Anticipation and deterrence alone are not enough to cope with 21st-century challenges and uncertainties. We deter threats we can anticipate, based on the known capabilities and suspected intentions of potential adversaries. Many (though by no means all) new types of threats will be unknown, unanticipated, or undeterrable. Thus, we must build resilience, which means acting now to improve our ability later to withstand the consequences of both probable and unpredictable threats. Resilience is much more than disaster preparation; it involves recognizing a continuum of risk and building up core capabilities that enhance the nation's ability to recover from a disaster or shock, to "build back better" than what was destroyed.

The United States needs a comprehensive strategy that includes investments in anticipation (to manage risk, allocate resources wisely, avoid or minimize shocks, prevent conflicts, and prepare for foreseeable outcomes), deterrence (to prevent politico-military competition from escalating into war), and resilience (to withstand shocks that were not prevented or perhaps preventable). Anticipation, deterrence, and resilience can help provide stability and improve policymakers' ability to manage in uncertain times. Elements of all components exist inside the U.S. government today, but they are not intellectually or bureaucratically integrated. The agencies and officials responsible for anticipation are decidedly not those working on deterrence or resilience.

Working on the Anticipation Equation

Anticipation is a hard problem that arguably is getting harder. Current complex and turbulent conditions make intelligence analysis even more difficult than it was during the relative stasis of the Cold War. Unsettled times make discontinuities more likely, yet tipping points are notoriously hard to identify or anticipate. It is difficult to strike the right balance between too many warnings, some of which are false alarms (or "crying wolf"), and not enough warning, which leads to surprise.

The attention of top policymakers is a finite commodity facing increasing demand. "Bandwidth" can be as critical a resource as money. Policymakers understand the need for good intelligence, but say they are too preoccupied with known problems of immediate consequence to focus on nonspecific intelligence warnings that are not "actionable." Intelligence professionals say that policymakers are too busy to address longer-term problems before they develop into crises—thereby increasing the likelihood that crises will occur. Both sides of the intelligence-action relationship are well aware that the political consequences of every bad outcome are all the more acute due to today's round-the-clock media climate, which can make every urgent global crisis appear to be an American problem, and the increasing polarization of U.S. foreign policy, in which any presidential decision that can be depicted as a mistake will be so portrayed.

Anticipation is not intelligence. Anticipation can begin when the intelligence community collects potentially useful information, analyzes it, decides what is important or actionable, and delivers that assessment to policymakers. Or it can begin when a senior policymaker asks a question. For example, when the President asks a question that his morning briefer cannot answer, the intelligence community spends a great deal of time on that question. The policymakers' challenge is to use intelligence, and a wide range of unclassified information, to anticipate likely and possible developments and act upon this to U.S.

advantage. For example, accurate intelligence can help one understand the intentions of an enemy and thereby deter, outwit, or defeat him. But that requires foresight.

Foresight involves asking not just what *will* happen, but what *might* happen, and what *could* happen as a consequence if it did. It is not enough for policymakers to imagine; they also must act, usually on imprecise information. And as senior policymakers quickly learn, preparation is even harder than foresight, for at least three reasons.

First, preparation requires accepting that something could happen—a cognitive hurdle, especially if the imagined event is deemed improbable, if not irrelevant or preposterous. The idea that terrorists could destroy the World Trade Center and change the course of U.S. history armed with only box cutters might have been judged preposterous—until it happened. But an event need not be preposterous to be rejected, merely out of sync with a favored perceptual model.[12] Having accepted that something might happen, policymakers must then prioritize, deciding how much effort and how many resources to invest in preparing for the possibility it will happen, and what kind of immediate response is warranted. For example, having foreseen climate change, we may need to prepare for more, and more intense, storms, but we also need to confront the issue of reducing greenhouse gas emissions now. Both actions count as anticipation. Foresight without action leads to regret.

The second hurdle to preparation is assessing risk. In traditional policymaking, *risk* is likelihood multiplied by consequence of a given event. When likelihood is almost impossible to foresee, this approach to assessing risk flounders. As Goldin and Mariathasan note, risk is supposed to be quantifiable and predictable, while uncertainty arises from unidentified or unexpected threats.[13] However, "given the pace of change, the traditional concepts of risk have become increasingly inappropriate as a basis of modern global governance."[14]

Third, preparation requires taking action to prevent, shape, or improve one's ability to react to or withstand not just one possible out-

---

[12] Gompert, Binnendijk, and Lin, 2014, pp. 174–194.

[13] Goldin and Mariathasan, 2014, pp. 25–28.

[14] Goldin and Mariathasan, 2014, p. 27.

come, but multiple plausible outcomes. This presents a political and bureaucratic hurdle because policymakers understandably want to allocate their resources based on how probable a threat is, not how plausible it is. We will discuss methodological responses to this problem.

Given these hurdles, it is perhaps unsurprising that the foresight and preparations that are required for anticipation remain elusive. But anticipation becomes more possible when one has a strategy. One successful illustration was the Carter Doctrine and subsequent U.S. efforts to keep the Soviet Union out of the Middle East.[15] In 1979, following the surprising Soviet invasion of Afghanistan and the fall of the Shah in Iran, U.S. officials began to worry that the Soviet Union, whose troops were within 300 miles of the strait of Hormuz, could invade Iran and push south to control the Persian Gulf, seize Saudi Arabian oil fields and cut off oil supplies to West. (There was later concern that Iran, which had recently been supplied with top-of-the-line U.S. weaponry and was controlled by a radical anti-American regime, would attack Saudi oil fields.)

It is unknown whether the Soviets ever, in fact, had such a plan—many analysts doubt it—but by the summer of 1980, the Central Intelligence Agency (CIA) believed the Soviet Union was "taking steps to strengthen the ability of its forces to invade Iran, should Soviet leaders so decide."[16] The Joint Chiefs of Staff concluded that the United States could not stop them.[17]

Based on the analysis that Moscow could succeed, with devastating consequences to the oil-dependent West if it did, the United States set about strengthening its deterrence. Specifically, it sought to deter the Soviet Union from moving against Iran or Saudi Arabia, either directly or through proxies, and threatening Gulf oil supplies.

---

[15] The doctrine was named after the declaration by President Jimmy Carter, in his January 23, 1980, State of the Union address, that "an attempt by any outside force to gain control of the Persian Gulf region will be regarded as an assault on the vital interests of the United States of America, and such an assault will be repelled by any means necessary, including military force." Jimmy Carter, *The State of the Union Address Delivered Before a Joint Session of the Congress*, January 23, 1980. Online by Gerhard Peters and John T. Woolley, American Presidency Project.

[16] See Henrik Bliddal, *Reforming Military Command Arrangements: The Case of the Rapid Deployment Joint Task Force*, Carlisle, Pa.: U.S. Army War College, 2011, p. 47.

[17] Bliddal, 2011, p. 49.

The State Department snapped into action, negotiating the rights to place or expand U.S. military bases on Diego Garcia, an atoll in the Indian Ocean, and in Somalia, Kenya, and Oman. DoD established the Rapid Deployment Joint Task Force, which later became the U.S. Central Command.[18] It also worked with the Gulf States to develop military infrastructure throughout the Gulf region. Thus began a 36-year U.S. military involvement in the Middle East. Whatever the long-term merits of that policy,[19] the Soviet Union did not enter the Persian Gulf and the oil has kept flowing through the Strait of Hormuz to the United States and other buyers ever since.

When it comes to disasters, the United States knows how to prepare. Following the debacle of Hurricane Katrina in 2005, the federal and state governments were better prepared for Superstorm Sandy in 2013. Meteorologists forecast Sandy in time to warn the public. Authorities took appropriate steps to mitigate the damage, and the federal, state, and local responses following the storm were widely praised, although some significant disruptions to liquid fuel supplies still occurred.[20] But U.S. health agencies were unprepared to help African nations deal with the rapid spread of Ebola in West Africa, although the possibility (though not the probability) of such a pandemic had been widely foreseen. And there are multiple warnings from the scientific community that the United States remains ill-prepared and is underinvesting in resilience to deal with the effects of a climate that could change much sooner than we think.

The United States has arguably been too slow to "get ahead of the game" on climate change not only because of the cost of doing so, but also because the threat poses all of the cognitive, methodological, and

---

[18] Joe Stork, "The Carter Doctrine and U.S. Bases in the Middle East," *Middle East Report*, Vol. 10, September/October 1980.

[19] For the argument against U.S. forward-basing policy see, for example, Andrew Bacevich, who argues that the Carter Doctrine was a seminal mistake. Andrew J. Bacevich, *America's War for the Greater Middle East: A Military History*, New York: Penguin Random House, 2016.

[20] Center on Global Energy Policy, "Summary of the Roundtable on the Resilience of the Liquid Fuel Supply Chain in the New York City Area," Columbia University, May 28, 2014; New York City Special Initiative for Rebuilding and Resiliency, *PlaNYC: A Stronger, More Resilient New York*, June 2013.

political hurdles already described. Harvard University cognitive psychologist Daniel Gilbert identified the human limitations of thinking about nonhuman challenges a decade ago:

> Global warming lacks a mustache. No, really. We are social mammals whose brains are highly specialized for thinking about others. Understanding what others are up to . . . has been so crucial to the survival of our species that our brains have developed an obsession with all things human. . . . Global warming isn't trying to kill us, and that's a shame. If climate change had been visited on us by a brutal dictator or an evil empire, the war on warming would be this nation's top priority.[21]

It is the job of government to factor in such human foibles and put in place systems to make good decisions despite them. But that requires a deliberative long-term planning process. The U.S. military is well prepared to deal with threats it foresees because it has such a well-honed, formal defense planning process. Civilian decisionmakers do not, but they may still benefit from working on a number of elements of the anticipation equation.

The first element is to better understand surprise. The larger the discrepancy between one's subjective (but distorted) model of reality and the objective reality, the greater the surprise. Decisionmakers with especially high confidence in their distorted models of reality were found to have the strongest tendency to resist information that would correct their misconceptions.[22]

Next, policymakers will want to improve intelligence collection and analysis, and thus foresight. Anticipation also requires improving policymaker capacity to ask for and absorb intelligence about events that are deemed low probability but would be highly consequential if they happened—precisely those most likely to produce strategic surprise and thus demand anticipatory action.[23] This may require creating

---

[21] Daniel Gilbert, "If Only Gay Sex Caused Global Warming," *Los Angeles Times*, July 2, 2006.

[22] Gompert, Binnendijk, and Lin, 2014, p. 181.

[23] James A. Dewar, *Assumption-Based Planning—A Tool for Reducing Avoidable Surprises*, Cambridge: Cambridge University Press, 2002.

bureaucratic mechanisms to combat cognitive bias and integrate thinking about low-probability/high-consequence problems into the policymaking process. It also requires distinguishing between long-term problems and low-urgency problems Policymakers, being human, tend to equate a long-term problem with lack of importance or urgency. In fact, some very long-term problems may nevertheless require urgent action to forestall bad outcomes—even though those outcomes may not occur for some time.

Policymakers may also consider using decision-aids. This might include more extensive use of "red-teaming" outside DoD and of human plus computer collaboration.[24] Scientists have learned that teaming up humans and computers—each doing what he, she, or it does best—produces better results than humans or computers working alone.[25] It turns out that merely competent human chess players teamed with computers can beat both computers and human grand masters. This insight has led to a new generation of advanced computer-human decisionmaking tools. The basic idea is to use computers in an iterative process of discovery together with humans to stress-test plans against a very large number of plausible paths into the future.[26] The information is presented in the form of scenarios, with the attendant sense of plausibility (not probability) that allows them to be more easily accepted even if they run contrary to accepted wisdom.[27] Because these scenarios emerge from an

---

[24] Micah Zenko, *Red Team: How to Succeed by Thinking Like the Enemy*, New York: Basic Books, 2015.

[25] Clive Thompson, *Smarter Than You Think: How Technology Is Changing Our Minds for the Better*, New York: Penguin, 2013.

[26] Robert J. Lempert, Steven W. Popper, and Steven C. Bankes, *Shaping the Next One Hundred Years: New Methods for Quantitative, Long-Term Policy Analysis*, Santa Monica, Calif.: RAND Corporation, MR-1626-RPC, 2003; Robert J. Lempert, "Can Scenarios Help Policymakers Be Both Bold and Careful?" in Francis Fukuyama, ed., *Blindside: How to Anticipate Forcing Events and Wild Cards in Global Politics*, Washington, D.C.: Brookings Institution Press, 2007; Robert J. Lempert, Drake Warren, Ryan Henry, Robert W. Button, Jonathan Klenk, and Kate Giglio, *Defense Resource Planning Under Uncertainty: An Application of Robust Decision Making to Munitions Mix Planning*, Santa Monica, Calif.: RAND Corporation, RR-1112-OSD, 2016.

[27] Robert J. Lempert, "Scenarios That Illuminate Vulnerabilities and Robust Responses," *Climatic Change*, Vol. 117, No. 4, 2013.

analysis aimed at identifying vulnerabilities of proposed strategies, they are harder to reject. They can inform discussion among officials who need not understand modeling—or computers, for that matter.

A final element of anticipation is identifying opportunity. Foresight about positive developments can be as valuable as warnings of danger—if the information is properly used. Various elements of the U.S. government are already engaged in some of these efforts. Nevertheless, new developments and best practices are not disseminated or adopted across government in a speedy fashion. The next President has the opportunity to demand fresh, high-level attention to improving the intelligence community's ability to foresee problems and the National Security Council's ability to decide what to do and to do it promptly. This requires a deeper understanding of surprise and the constraints on intelligence collection and analysis.

Understanding Surprise

The U.S. intelligence community comprises 16 federal agencies and even includes an "Office for Anticipating Surprise."[28] Nonetheless, the United States is all too frequently surprised.

Some of these surprises can be attributed to intelligence failures—that is, failures of fact or of foresight when the intelligence community is actively collecting information in a given area. Examples include the failure to foresee the collapse of the Soviet Union at a time when it was the leading U.S. intelligence target, or the failure to anticipate Russia's entry into the Syrian civil war while the intelligence community had its sights trained on ISIS. Intelligence failures can occur when the intelligence community points its lens in the wrong direction or zooms in too close to see broader trends. The failure to anticipate the Arab Spring was due in part to U.S. policymakers' focus on counterterrorism cooperation with the intelligence services in those countries, not on the stability of the regimes themselves.

---

[28] The office is part of the Intelligence Advanced Research Projects Activity under the Office of the Director of National Intelligence. See Intelligence Advanced Research Projects Activity, "Decoding the Future for National Security," undated-a, and "Office for Anticipating Surprise and the 'Super Forecasters,'" web pages, undated-b.

Some surprises are the result of policymakers' failure to translate analytic warnings into prompt or effective policy responses. (This could be due to failures of factual indicators, cognition, or appropriate action.) This seems to have been the case in the 9/11 attacks, which were mentioned in the August 6, 2001, presidential daily briefing.[29] Gompert, Binnendijk, and Lin note that poor *use* of information (not *lack* of information) is a primary cause of strategic blunders.[30]

Some "surprises" are developments that were well known to be possible if not probable, but deemed unlikely to happen so soon (that is, failures of anticipation, due to faulty assumptions). This type of temporal surprise was epitomized by the death of an Arctic explorer who died by falling through the unexpectedly thin ice whose melting he had set out to measure.[31] The United States was caught flat-footed by the overthrow of Egyptian President Hosni Mubarak in 2010—even though the CIA had warned of the risk. As former CIA Deputy Director Michael Morell wrote, "for a number of years CIA analysts had been warning about powerful pressures in the Arab world. In one piece after another they told policymakers that, without significant policy change in the Arab world, the status quo would not stand."[32] In some

[29] The possibility of an attack was repeatedly discussed from May to August 2001, but national security adviser Condoleezza Rice disputed that the item in the presidential daily brief, titled "Bin Laden Determined to Strike the U.S.," constituted a call to action, as the Federal Bureau of Investigation was actively hunting down al Qaeda cells in the United States. See Thomas S. Blanton, "The President's Daily Brief," Washington, D.C.: National Security Archive, April 12, 2004.

[30] "First, correct information may exist but may not penetrate bureaucratic barriers to reach the designated policymaker, thus having little influence on the final decisionmaking. Second, even if policymakers had valuable information, their cognitive biases may cause them to disregard the information or use the information incorrectly." Gompert, Binnendijk, and Lin, 2014, pp. 27–28.

[31] Gayathri Vaidyanathan,"Thinning Arctic Ice Proves Deadly for Veteran Explorers: The Polar Region's New Normal Seems to have Contributed to the Deaths of Two Experienced Arctic Researchers," *Scientific American*, May 6, 2015.

[32] Michael Morrell and Bill Harlow, *The Great War of Our Time: The CIA's Fight Against Terrorism from al Qa'ida to ISIS*, New York: Twelve, 2015, pp. 178.

cases, blunders occur because "the models of reality that guided decisionmaking [are] defective."[33]

Failure to consider the effects of U.S. behavior itself is a more important factor than we might imagine. By law, the intelligence community collects intelligence on foreign targets, not its own government, and for good reason. Yet, as became clear after 9/11, the United States is an immensely powerful actor whose decisions can significantly alter the course of many (but not all) events. Deeper consideration of unintended consequences is essential.

Intelligence Collection

Two other structural constraints on intelligence-gathering may hinder anticipation. The first is the requirement that the intelligence community focus most collection on the policymakers' highest priorities. At present, that means a concentration of resources to support counterterrorism operations in Afghanistan, Iraq, and Syria, at the expense of collecting on other topics. The intelligence community takes seriously its mission to serve "the customer"—the policymakers who consume their products. In wartime, customers do not want to compromise the mission by diverting collection efforts to secondary or tertiary priorities. Yet surprises—malign and otherwise—are, by definition, most likely to emanate from places and problems that are not under scrutiny. The intelligence community would benefit from a more inquisitive customer, one that is persistent in asking what might lurk unseen around the corner.

The second constraint is the inherent difficulty in collecting information about the intentions of leadership in foreign countries and nonstate groups. Potential U.S. adversaries are mainly authoritarian, closed organizations. The top leadership circles in Moscow, Tehran, Beijing, and Pyongyang tend to be small, secretive, and, in the wake of disclosures about National Security Agency interceptions abroad, perhaps even more cautious than before. The Russian foreign ministry is reported to be using typewriters to prevent digital surveillance,[34] even

---

[33] Gompert, Binnendijk, and Lin, 2014, p. 176.

[34] For example, in the wake of the Wikileaks disclosures, the Russian Foreign Ministry reportedly placed an order for manual typewriters. J. Dana Stuster, "Russian Security Now

as Russians reportedly hacked into U.S. electoral databases. Osama bin Laden sent messages by courier to avoid detection. Learning the intentions of secretive leaders confers strategic advantage.

Intelligence Analysis

Obviously, intelligence that is not collected is not analyzed. Moreover, as with intelligence collection, resources for intelligence analysis are also allocated based on customer demand and priorities. Policymakers understandably have a strong bias toward issues they deem to be of immediate importance, and less interest in longer-range problems of questionable probability. Inevitably, the need to inform immediate decisions and operations crowds out analytic resources for systematic thinking about long-term, over-the-horizon possibilities.

Policymakers and intelligence professionals understand full well that humans tend to conflate the urgent with the important and the unknown with the unlikely.[35] Nonetheless, they have no established mechanism that combines intelligence analysis and discussion of the perceived probability of a development with its likely consequences. Given the current environment of uncertainty on a large number of emerging or probable issues, however, the policy community would benefit from a fourth, formal mechanism that could alert policymakers to developments that would be particularly consequential, even if they are assumed to be unlikely. Such a mechanism could also be useful for calling attention to slow-moving problems that nonetheless require immediate policymaker attention. The challenge, of course, is to do so without falling prey to the "boy who cried wolf" syndrome.

"Bandwidth" and "Demand Signal" Problems

Senior policymakers consistently complain that they have no time to think and that the nature of their jobs is increasingly reactive.[36] A number of different institutional and bureaucratic reforms have been

---

Using Typewriters to Thwart the NSA," *Foreign Policy*, July 11, 2013.

[35] Welton Chang, "Five Psychological Biases That Plague Policymakers and Military Leaders," *The Policy*, February 8, 2016.

[36] Smith, 2016.

suggested,[37] but none are likely to succeed without a President seized with the desire to improve the intelligence and advice he receives.

Any organizational remedy should address two problems. One is creating a demand signal from "the customer" (senior officials, or, better yet, the President) to the intelligence community for over-the-horizon thinking and deeper consideration of the unexpected. The intelligence community responds well to customer demand, as opposed to pursuing what it thinks policymakers ought to be interested in or could want to know. Policymakers are more likely to pay attention if they believe they asked the question the intelligence community is answering. The second solution requires fencing off policymaker time to consider anticipating responses to the intelligence thus generated. Again, this requires the President to insist that senior officials muster the "bandwidth"—time and attention—to identify nascent opportunities and ponder how to bring them to fruition. Anticipation is as much about pursuing opportunities as it is about dealing with threats.

(Re)Learning What it Takes to Deter

As we have seen, many of the assumptions that animated U.S. foreign policy in the post–Cold War era no longer apply. The instruments of warfare and the character of economic competition have evolved. So have the strategic orientations of Russia and China. These changes require a re-examination, and perhaps a (re)learning, of what it means to deter. Deterrence is a highly specific form of anticipation that must be updated for 21st-century realities.

A Short History of Deterrence

One of the hallmark strategies for managing the Cold War competition with the Soviet Union was deterring the use of force, or as Glenn Snyder expressed it, "discouraging the enemy from taking military

---

[37] Among the suggestions are, for example, a long-term planning office located in the White House, the National Intelligence Council, or the State Department's Policy Planning Bureau, or an independent body reporting to the President.

action by posing for him the prospect of cost and risk outweighing his prospective gain."[38] Enormous effort was devoted to understanding what would be required to deter Soviet aggression and under what circumstances. It involved military planners, large segments of the policy and intelligence communities, including national leaders, and important elements of the academic community.[39]

Deterrence is simple in concept but complicated in execution. The reason, as the late Thomas Schelling pointed out in his seminal work, *Arms and Influence*, is that ". . . deterrence is about intentions—not just estimating enemy intentions but *influencing* them" (emphasis in original).[40] Influencing intentions is an extraordinarily complicated matter.

Early thinking on deterrence made a distinction between punishment and denial. Threats of punishment were intended to convince an adversary that an act of aggression would be subject to a prohibitive response—that the United States could hold at risk something of greater value than that which an adversary was seeking. The advent of nuclear weapons (which provided the option of an overwhelming and devastating response), along with clear U.S. nuclear superiority in the early years of the Cold War, made the threat of punishment—as embodied in the Dwight Eisenhower policy of massive retaliation—all the more credible. As U.S. nuclear superiority eroded in the late 1950s and early 1960s, many came to recognize that not all threats of punishment were sufficiently credible, even when the stakes were high for both parties.

Consequently, from the early 1960s onward, the United States placed increasing emphasis on deterring the adversary—the Soviet Union and its Warsaw Pact allies—not solely through the prospect

---

[38] See Glenn Snyder, *Deterrence and Defense*, Princeton, N.J.: Princeton University Press, 1961, p. 3.

[39] Some of the earliest work on deterrence was done at RAND. See Bernard Brodie, "The Anatomy of Deterrence," Research Memorandum, U.S. Air Force Project RAND, RM-2218, 1958.

[40] Thomas C. Schelling, *Arms and Influence*, Princeton, N.J.: Princeton University Press, 1966, p. 35.

of punishment through nuclear retaliation, but by posing the credible prospect that U.S. and NATO *conventional forces* would be sufficient to prevent the adversary from achieving its aims through military aggression. The strategy was premised upon denying the adversary its aims by using forward-deployed forces, backed by well-exercised reinforcements. Over time, these conventional forces were more numerous and capable—and thus far more credible—than in the 1950s and early 1960s. Should they fail to deter or defeat Soviet military aims, then theater-based "tactical" nuclear forces, and ultimately "strategic" nuclear forces based in the United States promised punishment.[41] In this way, denial strategies combined with detailed escalation plans, including nuclear escalation plans premised on increasing levels of punishment, became a hallmark of Cold War planning.

During the Cold War, the United States had the advantage of focusing its deterrent efforts largely on one rival, the Soviet Union, and on specific actions—for example, an attack by Warsaw Pact forces on NATO Europe. Even then, the undertaking was vast and complex. The United States invested great amounts of time, effort, and resources in seeking to understand the Soviet Union—the motivations and intentions of Soviet leaders, how culture and history might condition Soviet decisionmaking (especially in time of conflict), and, of course, on gaining a detailed understanding of Soviet military capabilities and supporting doctrine. As geopolitical, regional, and technological circumstances changed, which they frequently did, the United States adapted its deterrent posture.

Much of the serious American thinking on deterrence, both nuclear and conventional, ceased with the collapse of the Soviet Union. The new Russian Federation was a partner, and a democratic one, at that.[42] U.S. troops in Europe were cut from almost 260,000 in 1991 to

---

[41] At the level of policy, America's purposeful adoption and integration of a combination of denial and punishment strategies—referred to by a variety of labels ("flexible response" being perhaps the best known among them)—marked a profound and, in time, increasingly pronounced shift away from the "massive retaliation" policy of the early Cold War.

[42] See Jeffrey D. McCausland, *The Post 'Post-Cold War Era' in Europe*, Carlisle, Pa.: Strategic Studies Institute, U.S. Army War College, April 14, 2014.

approximately 52,500 in early 2016,[43] and successive rounds of negotiations cut the two countries' nuclear arsenals by more than two-thirds.[44] On the Korean Peninsula, U.S. forces continued to maintain a largely deterrent posture in the face of the unrelenting and now nuclear threat posed by the hostile regime in Pyongyang. When the Cold War came to an end, the United States did not abandon either the concept or the capabilities of deterrence, but many of the people who had spent much of their professional lives thinking about how to deter Soviet aggression moved on to other pursuits.

Interest in deterrence faded further after the 9/11 attacks, when the Bush administration argued that traditional deterrence would not work against terrorists and "rogue states."[45]

A deterrence strategy is both about actors and actions. One actor does not deter another in a general sense. Rather, one actor seeks to deter another actor from taking a particular action. This concept needs to be at the forefront of policy thinking as the country contemplates how to preserve the peace in a repolarizing world. Recent

[43] During the height of the Cold War, there were more than half a million U.S. personnel assigned in the European theater. As of February 2016, around 52,500 personnel were in direct support of U.S. European Command missions, while another 9,500 personnel supported the missions of other organizations, such as U.S. Africa Command and U.S. Transportation Command. See Gen. Philip Breedlove, *U.S. European Command Posture Statement 2016*, U.S. European Command, Stuttgart, Germany, February 25, 2016.

[44] Together, the Strategic Arms Reductions Treaty (START, entry into force 1994), the Strategic Offensive Reductions Treaty (SORT, entry into force 2003), and the New START agreement (entry into force 2011) resulted in deep reductions in the numbers of deployed strategic warheads and deployed delivery vehicles by each country. In addition, while considerable uncertainty exists about the current state of Russia's tactical nuclear forces, the Presidential Nuclear Initiatives announced by President George H. W. Bush on September 27, 1991, and by Soviet leader Mikhail Gorbachev on October 5, 1991, just seven weeks prior to the official dissolution of the Soviet Union, resulted in the elimination of all U.S. nuclear artillery shells and short-range nuclear ballistic missile warheads and the removal of all nonstrategic nuclear warheads from U.S. surface ships, attack submarines, and land-based naval aircraft.

[45] The Bush administration's National Security Strategy, released in September 2002, one year after 9/11, stated, "Traditional concepts of deterrence will not work against a terrorist enemy whose avowed tactics are wanton destruction and the targeting of innocents; whose so-called soldiers seek martyrdom in death and whose most potent protection is statelessness." White House, *National Security Strategy of the United States of America*, Washington, D.C., September 2002, p. 15.

Russian and Chinese behaviors, to say nothing of threats elsewhere, call for a reimagining of deterrence for a more complex environment. Moreover, a one-size-fits-all deterrence strategy will not work against the multiplicity of actors and the multiplicity of challenges the United States will face in the decades ahead.

Conventional Deterrence and the Problem of Asymmetry

To meet the challenges that will likely face the United States in the coming years, we argue that the nation's immediate focus must be to (a) emphasize conventional means of deterrence relative to nuclear means; (b) prepare to deter across multiple domains; and (c) tailor deterrent capabilities and messages to influence different potential adversaries.

In the short term, the problem is not deterring nuclear adversaries from directly attacking the United States; there is no lack of credibility regarding U.S. ability or willingness to use nuclear weapons in retaliation for nuclear strikes on the homeland. Rather, the core problem is deterring adversaries from undertaking conventional forms of aggression in regional conflicts for limited objectives. There is concern among some U.S. allies about the willingness of the United States to honor its security commitments, including, in extremis, its nuclear deterrence commitments, in the event that the allies are the victims of aggression. The U.S. task lies in extending its deterrent posture to allies and partners in a manner that is simultaneously as credible as it is unmistakably capable.

This is not a new problem, but rather an old one that has reemerged. Throughout the Cold War, U.S. allies harbored doubts about the credibility of U.S. commitments and the willingness to use force, including nuclear weapons, should their territory be threatened. In a memorable Cold War exchange, France's Charles de Gaulle is said to have asked President John F. Kennedy, "Would the United States be willing to trade New York for Paris?"[46] Although de Gaulle's quip prompted a variety of glib rejoinders at the time, his question was one that worried allies

[46] John F. Kennedy and Charles de Gaulle, "Memorandum of Conversation," Paris, May 31, 1961, posted online in U.S. Department of State, Office of the Historian, *Foreign Relations of the United States, 1961–1963*, Vol. XIV, *Berlin Crisis, 1961–1962*.

throughout the Cold War: Would the United States be prepared to invite attack upon itself in coming to the defense of a friend? Would Washington be as committed to the defense of its allies as would-be adversaries are in threatening them? After all, by threatening to use nuclear weapons in defense of its allies, the United States was opening itself to nuclear retaliation; it was, in a very real sense, expressing a willingness to assume the risk that an American city would suffer a nuclear strike in retaliation for nuclear use undertaken in defense of a European city.

De Gaulle's question highlights an enduring problem that arises from what deterrence theorists refer to as the "asymmetry of stakes," those circumstances in which one party cares far more about the outcome than the other. If what is at stake for the adversary outweighs the U.S. interest in preventing them from intimidating, controlling, or seizing territory from American friends, then U.S. credibility will be questioned.

The perceived asymmetry of stakes could well be understood by an adversary as an asymmetry of will—with the United States lacking the will, though not necessarily the means, to mount a viable and effective response to aggression. Back to Schelling's earlier point: Deterrence is not just about *estimating* enemy intentions but *influencing* them.

Although today's U.S. allies and partners may be less outspoken than de Gaulle, they are likely to harbor similar doubts about the depth of the U.S. commitment to their defense, for here is what they observe: China and Russia are developing forces, doctrine, and weapons designed around presenting a *fait accompli*, to achieve quick victories against nearby states before U.S. forces can intervene in sufficient numbers to make a difference.[47] U.S. strategic thinkers and defense analysts share the concern that emboldened Chinese and Russian leaders might prove willing to run the risk of testing U.S. resolve by forcefully assuming control of territory and waters claimed by a U.S. ally in East Asia or, in Moscow's case, undertaking military aggression to reclaim one or more of the Baltic states. Their reasoning would almost

[47] For instance, China's strategy of "winning informationized local wars" envisions localized, short-duration, and high-intensity conflicts. See State Council Information Office of the People's Republic of China, *China's Military Strategy*, Beijing, May 2015.

certainly be based on the belief that they could achieve their principal objectives before the United States or its allies could meaningfully react to the situation. Presented with such a *fait accompli*, they might calculate that the United States and its allies would realize the enormity of the costs in lives and treasure that would be entailed in reversing the situation, including the distinct possibility that Russia or China might use nuclear weapons in a limited capacity, and conclude that the interests at stake are not so vital as to be worth the costs. In brief, the calculation that an adversary would be acting upon is that its stakes (or interests) in the conflict sufficiently outweigh those of the United States and its allies, and that sooner, rather than later, the United States will conclude that the game is not worth the candle.[48]

Deterrence is most likely to succeed when an aggressor believes it cannot achieve its objectives short of a prolonged conflict of uncertain outcome. As RAND colleague David Ochmanek has written, "The gold standard of deterrence . . . is to confront a potential aggressor with the credible prospect of failure."[49]

Were the United States to find itself in a situation in which Moscow or Beijing attacked a NATO or other U.S. treaty ally, and Washington did *not* respond with a viable conventional response or a credible threat of nuclear retaliation, its alliance commitments would be rendered meaningless. If Russia were to succeed in seizing the Baltics, for example, South Korea and Japan would have every incentive to develop more-independent security postures, possibly including nuclear weapons of their own. If China succeeded in grabbing Taiwan, then the newer members of NATO (every state that has acceded since the end of the

---

[48] That an adversary might use nuclear weapons with such an intent is not mere speculation, but has been the subject of Russian strategic military discussion in recent years. Recognizing the increasing standing such concepts are receiving from potential adversaries, the Department of Defense's 2014 Quadrennial Defense Review emphatically states, "Our nuclear deterrent is the ultimate protection against a nuclear attack on the United States, and through extended deterrence, it also serves to reassure our distant allies of their security against regional aggression. It also supports our ability to project power by communicating to potential nuclear-armed adversaries that they cannot escalate their way out of failed conventional aggression." Charles T. Hagel, *Quadrennial Defense Review*, Washington, D.C.: U.S. Department of Defense, March 2014, p. 12.

[49] Ochmanek et al., 2015.

Cold War) would have second thoughts about the value of U.S. security commitments, to say nothing of South Korea and Japan.

The obvious remedy is to strengthen U.S. *conventional military capabilities* to demonstrate that such attacks would not succeed. Yet there has been a notable erosion in conventional capabilities to deter both Russian aggression on its periphery and, increasingly, Chinese designs on the "first island chain."[50] The modernization of both militaries has been so rapid, and their targets so close to their own territory, that each has a "home field advantage" not held a decade ago. Moscow and Beijing understand the superiority of U.S. conventional military forces once the Americans have time to bring in reinforcements from afar. That is why they are planning for short wars.

Deterring Across Multiple Domains

The core deterrence imperative before us, therefore, is to deploy conventional forces capable of denying aggressors the ability to accomplish their goals quickly. This is followed closely by the imperative that U.S. forces be capable of deterring aggression across multiple domains to include land, sea, air, and, eventually, space and cyberspace.

Warfare across domains is not new. As Gartzke, Lindsay, and Nacht trenchantly observe, the Greeks responded to the abduction of Helen of Troy not by undertaking a campaign of counterabductions but by laying siege to Troy, and the British decided to destroy the French fleet in the Battle of the Nile rather than vainly attempting to undertake a ground campaign against Napoleon's army.[51] Likewise, the Japanese used kamikaze pilots against U.S. warships they could not sink by conventional means, and the United States used aircraft against Serb ground forces in Kosovo and to "soften up" Saddam Hussein's forces in Iraq and ISIS in Syria.

Two factors distinguish cross-domain conflict today and in the decades ahead. First, U.S. military forces and the U.S. economy are inextricably dependent on space and cyberspace—more dependent than

---

[50] Ochmanek et al., 2015.

[51] Erik Gartzke, Jon Lindsay, and Michael Nacht, "Cross-Domain Deterrence: Strategy in an Era of Complexity," International Studies Association Annual Meeting, Toronto, 2014, p. 14.

are less technologically advanced nations. Recognizing this military and economic vulnerability, U.S. defense leaders have invested considerable effort in understanding the effects of attacks on space and cyberspace. These efforts have included training in environments where space and cyber capabilities are not readily available or reliable. Without actual experience in combat, there is really no predicting how an actual attack might unfold and what the consequences would be, not only for our military forces but also critical national infrastructure, especially the electric power grid and water systems.[52] Second, given the importance of space and cyber operations, the likelihood that strikes on them could lead to major escalation in a conflict is seen as both high and increasing. Unlike in traditional domains, where escalation thresholds are at least conceptually, if not practically, understood and recognized by decisionmakers, there is very little understanding of what would constitute thresholds to escalation before or during conflict in space and cyberspace. The discussions between President Obama and Premier Xi Jinping during the 2015 U.S.–China summit, aimed at reducing the likelihood of cyber conflict, came about from the recognition on both sides that online clashes could escalate into conflict in the physical world.

Concerted thought will need to be given to what will constitute credible deterrence of aggression against U.S. space and cyber infrastructure and capabilities, while distinguishing cyber crime and espionage from cyber warfare. Although details have yet to be developed, it is likely that a policy of deterrence through denial of adversary objectives (e.g., via increased redundancy, passive security measures, the resilience to quickly restore services, and other defensive means) would be more promising and credible than a strategy based on deterrence by punishment. Given the outsize American reliance on space and cyber capabilities, adversaries may be only too pleased to engage in a "tit-for-tat" exchange with the United States.

---

[52] Martin C. Libicki, Lillian Ablon, and Tim Webb, *The Defender's Dilemma: Charting a Course Toward Cybersecurity*, Santa Monica, Calif.: RAND Corporation, RR-1024-JNI, 2015.

Tailoring U.S. Deterrence

Successful deterrence in the decade ahead requires U.S. decisionmakers and planners to understand and tailor deterrence policies, capabilities, and communications/messages (including declaratory policy) against at least three broad categories of actors and actions: (1) near-peer/major nuclear competitors (i.e., Russia and China) that threaten key allies and may be seeking to change the regional status quo, (2) nuclear-armed/ nuclear aspiring regional adversaries (i.e., North Korea and Iran) that similarly threaten key allies and may be seeking to change the status quo in key regions, and (3) terrorist networks seeking to obtain weapons of mass destruction.

Beyond this, detailed plans will need to be developed to achieve U.S. deterrence objectives. To be clear, the deterrence-by-denial strategy we advocate does not necessarily require maintaining forward-deployed forces around the globe. Competitive strategies can be effective, as can other approaches. Missile defense, for example, is a denial approach that does not necessarily require large numbers of forward-deployed forces.

Deterrence policy will only be as successful as the operational concepts developed and the associated capabilities postured to prevent specific actors from undertaking specific acts of aggression in specific contexts.[53] Knowing what capabilities and messages will prevent adversaries from undertaking a specific act of aggression—what will make them choose caution instead—will require an understanding of each of the different potential adversaries not dissimilar in scope and scale from that which was undertaken vis-à-vis the Soviet Union in the Cold War.

---

[53] The United States does not and cannot plan to "deter North Korea." Rather, it plans to deter North Korea from attacking South Korea via a missile barrage followed by a mechanized infantry attack, or it plans to deter North Korea from selling a nuclear weapon to a terrorist network, or to deter North Korea from attacking Japan with nuclear weapons via ballistic missile, and so on. For an incisive discussion from which this example was inspired, see M. Elaine Bunn, "Can Deterrence Be Tailored?" *Strategic Forum*, Institute for National Strategic Studies, National Defense University, No. 225, January 2007.

Building in Resilience

Resilience was a hallmark of Cold War–era planning. From the design of subway tunnels that were also intended to serve as radioactive fallout shelters to the creation of a government relocation facility to accommodate the House and Senate in the event of a nuclear war, the United States proceeded from the assumption that deterrence might fail. Like deterrence strategy, much of this planning was abandoned as the Cold War ended. Yet in the face of new challenges, it is possible that anticipation and deterrence will sometimes fail. Therefore, increasing resilience against a range of possible new threats will be just as important.

Resilience is a broad concept that cuts across all elements of U.S. strategy. It can be defined as "the capacity of any entity—an individual, a community, an organization, or a natural system—to prepare for disruptions, to recover from shocks and stresses, and to adapt and grow from a disruptive experience."[54] The nation's military posture and forces—land, sea, air, cyberspace, and space—must be resilient. That concept is well developed and goes far beyond the traditional military notion of "hardening the target" to include physical and psychological resilience of service members and their families.[55] The U.S. alliance structure ought to be resilient, so that if one ally falters, others can come to the rescue. The nation's energy grid and its communications infrastructure must be resilient, as must its financial system.[56] Businesses need to be resilient to handle disruptions to global supply chains, which improve their efficiency but also expose them to more risk.

---

[54] Judith Rodin, *The Resilience Dividend: Being Strong in a World Where Things Go Wrong*, New York: Public Affairs, 2014, p. 3.

[55] See, for example, Lisa S. Meredith et al., *Promoting Psychological Resilience in the U.S. Military*, Santa Monica, Calif.: RAND Corporation, MG-996-OSD, 2011; Gary McLeod et al., *Enhancing Space Resilience Through Non-Materiel Means*, Santa Monica, Calif.: RAND Corporation, RR-1067-AF, 2016.

[56] See, for example, Henry H. Willis and Kathleen Loa, *Measuring the Resilience of Energy Distribution Systems*, Santa Monica, Calif.: RAND Corporation, RR-883-DOE, 2015. For a look at systemic risks in the financial system and their management, see Goldin and Mariathasan, 2014.

At whatever scale disaster strikes, communities need to be resilient, too, as the first line of emergency responders and those closest to the people.[57] Community resilience has overlapping physical, psychological, and sociological dimensions. It can mean building physical infrastructure that is resistant to damage by natural disaster or terrorist attack, or improving capabilities that would allow for rapid rebuilding of what may be destroyed. It can also mean the community's capacity to bounce back in response to a disaster—manmade or natural—or even "bounce forward," emerging from accidents or traumas better than before. The idea of community resilience embodies the notion of government and civil society working together to assess vulnerabilities and then draw on the necessary assets to competently and swiftly cope with the immediate aftermath of disaster, thereby ensuring quick recovery. Research conducted following Hurricane Katrina has documented the unevenness among different segments of a community in how they recover—or do not recover—from trauma.[58]

Communities under stress from chronic poverty or high levels of violence, for example, are less likely to bounce back quickly, whereas strong, close-knit communities may be able to grow rapidly after a disaster, a phenomenon known as "posttraumatic growth." *Resilience* in this context means not only the ability of a community to deliver relief or relocate displaced people but also the capacity to leverage public and private funding to reconstitute public services and rebuild. Emergency preparedness and resilience capabilities can be developed, placing communities in the best position to respond and recover more effectively,

---

[57] See RAND Corporation, "Community Resilience," web page, undated-a; RAND Corporation, "Infrastructure Resilience and Environmental Policy," web page, undated-b; RAND Corporation, "Resilience in Action," web page, undated-c.

[58] The World Bank notes an upward trend in losses from natural disasters, estimating total losses from 1980 to 2012 at $3.8 trillion. Disaster is considered a structural impediment to sustainable development. See World Bank, "Managing Disaster Risks for Resilient Development," April 10, 2014. One study of disasters in Mexican municipalities found they set development back by the equivalent of two years. Eduardo Rodriguez-Oreggia et al., "Natural Disasters, Human Development and Poverty at the Municipal Level in Mexico," *Journal of Development Studies*, Vol. 49, No. 3, March 2013; Eduardo Rodriguez-Oreggia, *The Impact of Natural Disasters on Human Development and Poverty at the Municipal Level in Mexico*, Cambridge, Mass.: Center for International Development, Harvard University, Working Paper 43, January 2010.

but these capabilities are difficult to test in advance, particularly in the face of cascading disasters.

National efforts to increase community resilience to natural disasters, including climate change, have been under way at most major federal departments, including Health and Human Services, Homeland Security, and Justice.[59] Physical resilience measures may include seawalls and levees, storm surge barriers and dunes, rezoning or changes in land use, relocating vulnerable people to higher or safer ground, or improving building codes, as well as strengthening the human bonds within a community. Federal policies now increasingly focus on the vital role of local leadership, with cities taking much of the lead in developing resilience programs and strategies to respond to a range of natural, social, economic, and security stresses, such as countering violent extremism. For example, Secretary of State Kerry has stressed the need for local communities to build resilience against Islamist radicalization by addressing youth alienation and disaffection.[60]

From the standpoint of national strategy, active steps to improve resilience in many aspects of government planning and preparedness will be particularly important in coming decades, given the increasing recognition that the more globalized and complex our societies, the more exposed they are to systemic global risk.[61] Of particular concern is the lack of preparedness for next-generation cascading disasters (so-called "global network disasters" or "network of networks" disasters). As a disaster prevention specialist put it:

> The decadal trends and the best available science all clearly indicate that geophysical, meteorological, biological, technological

---

59 The Justice Department launched a "Strong Cities Network" to "strengthen community resilience against violent extremism." Office of Public Affairs, "Launch of Strong Cities Network to Strengthen Community Resilience against Violent Extremism," Washington, D.C.: Department of Justice, September 28, 2015.

60 John F. Kerry, "Remarks at the Strong Cities Network International Visitors Leadership Program for Municipal Leaders and Countering Violent Extremism Experts Event," Washington, D.C.: U.S. Department of State, March 1, 2016a.

61 The link between globalization and systemic risk, and the nature and consequences of those risks, are explored in depth by Goldin and Mariathasan, 2014, pp. 9–35, 198–220.

> and human induced disasters are increasing in intensity (also many in frequency), complexity (interconnected, synergistic and cascading), [and] uncertainty (future new events). Further, these multiplying risk factors are interacting with an ever more complex set of physical, social, economic and environmental vulnerabilities at rates that nations, societies and commerce are ill-prepared to deal with in terms of "gaps" in existing governance and institutional capacities.[62]

Scholarly and policy work on the cascading disaster problem intensified after Japan's 2011 Fukushima Daiichi accident, in which an earthquake of a magnitude of 9.0, a strength that most Japanese scientists had not deemed credible,[63] triggered a tsunami that killed more than 18,000 people and, in turn, caused meltdowns at three nuclear reactors that had been considered among the safest in the world.[64] Japan's then–Prime Minister Naoto Kan later said he was considering evacuation of metropolitan Tokyo—population 50 million—because the safety margin that kept the damaged reactor from releasing far more radioactive material was "paper thin."

Although the earthquake and tsunami were beyond human control, a commission appointed by the Japanese Diet to investigate the nuclear accident found that the reactor meltdown could not be regarded as a natural disaster. The commission chairman concluded that, "It was a profoundly man-made disaster—that could and should have been foreseen and prevented. And its effects could have been miti-

---

[62] Mika Shimizu and Allen L. Clark, "Interconnected Risks, Cascading Disasters and Disaster Management Policy: A Gap Analysis," planet@risk, Vol. 3, No. 2, 2015.

[63] Peter Landers, "The Man Who Predicted the Tsunami," *Wall Street Journal*, April 9, 2011.

[64] According to the National Police Agency of Japan, as of September 2016, 15,894 people were killed in the 2011 disaster and 2,557 are still missing. National Centers for Environmental Information, "March 11, 2011, Japan Earthquake and Tsunami," web page, March 2015; Yukiua Amano, *The Fukushima Daiichi Accident: Report by the Director General*, Vienna: International Atomic Energy Agency, 2015.

gated by a more effective human response"[65]—in other words, better resilience planning.

The cognitive biases that prevented Japan's TEPCO utility officials and nuclear regulators from effective tsunami planning were not universally shared. Another Japanese nuclear plant owner not far from Fukushima accepted the possibility of a stronger tsunami and built a higher sea wall, which did protect its plant. TEPCO did not. A Japanese paleoseismologist who had studied the geologic evidence of previous earthquakes and tsunamis had predicted that the region was overdue for a giant wave, and his findings were discussed at a 2009 meeting of a committee on nuclear power plant safety, but the national government took no action.[66] The scientist, Masanobu Shishikura, had an appointment with the Fukushima government to discuss his research on March 23. The earthquake struck on March 11. The country's tsunami warning system worked, but many people who lived at the high-water mark from a 1960 tsunami did not bother to run for higher ground.[67] The Fukushima tragedy demonstrates that the need for preparation and preventive action may be urgent even when the perception of risk is not. Humans tend to value future benefits less than immediate ones, but this is a cognitive error. As Rockefeller Foundation President Judith Rodin notes,

> Yes, it may be easier to mobilize people and to martial [sic] resources in the wake of a crisis, but there are at least two problems with the reactive approach to resilience building. First, you may be forced into actions that might not necessarily be optimal—such as constructing a new building when an existing one could have been made more able to withstand a disruption.

---

[65] Hiroko Tabuchi, "Inquiry Declares Fukushima Crisis a Man-Made Disaster," *New York Times*, July 5, 2012.

[66] Landers, 2011.

[67] This finding is relevant to West Coast tsunami education efforts, which are attempting to train at-risk populations in where and how fast to evacuate. See Nathan J. Wood et al., "Community Clusters of Tsunami Vulnerability in the US Pacific Northwest," *Proceedings of the National Academy of Sciences*, Vol. 112, No. 17, March 15, 2015.

> Second, the cost is always higher to take action postcrisis than it is when relief and recovery are not part of the agenda.[68]

From the standpoint of national strategy, research by RAND and others has identified several key elements of resilience that can be encouraged by the federal government in its domestic and international planning. These steps assume that more-frequent or more-severe disasters or shocks could be the new norm.

- Prepare for a long recovery period, and recognize that another disaster may strike while a community is still recovering from the first. Preparation includes establishing distributed, networked communication and response systems that can offset the destruction of centralized governmental systems (for example, in terrorist or cyberattacks).
- Encourage a more collaborative style of leadership and management. Federal, state, and local officials will need to work across sectors, bureaucracies, and institutions, and build closer ties among government, business, and civil society. In the short run, coordination costs time and money, but could save both when disaster strikes. Management systems must not be so "lean and mean," or dependent on single supply chains for vital relief supplies, that they cannot ramp up in times of disaster.[69]
- Redesign policies, programs, infrastructure, and virtual networks to strengthen adaptability, recovery, and resilience capabilities. For example, fund utility and public works projects with resilience built in, such as green infrastructure or redundant electric transmission systems that provide back-up capabilities. Such projects should be designed to provide economic benefits under busi-

---

[68] Rodin, 2014, pp. 282–283.

[69] Goldin and Mariathasan note the need for "buffers, safety nets and emergency procedures." A prudent approach to risk management for government and companies is to "consider backup structures as more than 'dead capital' or redundant investments. Lean management can be beneficial when things go well. More spare capacity, however, is required in an increasingly interconnected, interdependent and riskier world." Goldin and Mariathasan, 2014, p. 207.

ness-as-usual conditions and be robust and resilient in times of disasters or other acute shocks.

Turbulent times require deliberate, adaptive strategies to prepare for change, anticipate the unexpected, and to translate foresight into action. Budgetary constraints must be balanced with rigorous efforts to evaluate and manage risk. Resilience thinking needs to be a central feature of any future planning effort.

Part Two
Strategic Choices

CHAPTER SIX

About Strategy—Grand or Otherwise

> . . . The realm of strategy is one of bargaining and persuasion as well as threats and pressure, psychological as well as physical effects, and words as well as deeds. This is why strategy is the central political art. It is about getting more out of a situation than the starting balance of power would suggest. It is the art of creating power.
>
> Lawrence Freedman[1]

Over the last few decades, there have been many efforts among academia and think tanks to develop a strategic concept or orientation appropriate to a time of dramatic change in world affairs (see the Appendix for details). No consensus has emerged from these efforts, but certain assumptions about America's role in the world are widely shared:[2] that we are living in an era of turbulence driven by transformative political, economic, and technological change; that the United States is not able to disconnect from the world or assume it is protected from global turmoil by the oceans; that the United States must continue to exercise some form of international leadership; that the country's resources for supporting international activism are limited;

[1] Lawrence Freedman, *Strategy: A History*, New York: Oxford University Press, 2013, p. xii.

[2] However, these elite views were not shared by a number of candidates in the 2016 presidential campaign.

and that policies should be designed to promote fundamental national values as well as interests.

Despite this general agreement, there are significant areas of disagreement: how much to emphasize domestic needs versus international challenges; the appropriate range of U.S. activism and leadership abroad; the burdens the United States should accept in meeting commitments to its allies; whether the postwar liberal consensus is breaking down or simply readapting to changing realities; and which challenges to U.S. security, economic interests, and national values should be given highest priority. These differences tend to be refracted through a political lens regardless of whether they are presented as "grand strategy."

At the highest level of conceptualization, a country's *grand strategy* has been characterized as a "conceptual center of gravity" for national policy making.[3] Articulating a grand strategy has become the basis for building public support for a country's foreign policies and defenses. During the Cold War, America's grand strategy for dealing with the Soviet threat developed—over time and through crises and vigorous debate—around the concepts of containment and deterrence.[4] Is it possible to have such a unified and concise strategic concept in today's world of myriad, complex, rapidly moving developments that ping policymakers' smart phones 24/7? Some argue that the issues are too myriad and complex, change too rapidly, and require responses too immediately to make it possible to adhere to any enduring strategic conceptualization.[5] They note that strategic planning has fallen out of favor in the corporate world because it is seen as an impediment to organizational agility. Others argue that the bureaucratic processes of

---

[3] Hal Brands, *What Good Is Grand Strategy? Power and Purpose in American Statecraft from Harry S. Truman to George W. Bush*, Ithaca, N.Y.: Cornell University Press, 2014, pp. 8–9.

[4] Among the voluminous literature on this subject, we have found the following especially helpful: Ernest R. May, ed., *American Cold War Strategy: Interpreting NSC 68*, Boston: St. Martin's Press, 1993; John Lewis Gaddis, *The Cold War: A New History*, New York: Penguin Press, 2005.

[5] "There is no principle that can provide strategic clarity or unity of purpose to [today's] diverse challenges. The Cold War was an aberration in having such a single goal. Americans will have to live without a replacement for containment . . ." Jessica Mathews, "What Foreign Policy for the US?" *New York Review of Books*, September 24, 2015, p. 44.

policy implementation work against strategic planning,[6] or even that the effort to conceive a "grand" strategy is not just a delusional effort but a dangerous policymaking "ritual."[7]

The American approach to foreign policy management is decidedly pragmatic and ad hoc. Our political culture does not incline to grand concepts. Better to deal with issues as they come along and on their own terms—an approach that comes naturally to the lawyers and business leaders who tend to occupy senior policymaking positions in the U.S. government. These professionals are trained to deal with issues in practical terms, case by case, within a legal or rules-based framework of decisionmaking. Indeed, at least one president, Clinton, is said to have rejected the notion that grand strategy was a useful concept.[8]

Critics counter that pragmatic and ad hoc approaches produce reactive, incoherent policies that are confusing to the public, as well as to allies and adversaries, and are not conducive to the pursuit of long-term national interests. At best, they lead to budgets that define strategies.[9] At worst, they lead to poor outcomes and a decline in national power.

Given these concerns and critiques, should the next administration try to develop a grand strategy or opt for a more humble, ad hoc, or flexible approach?

We conclude that a presidential formulation of an overarching statement of national purpose and priorities for engaging the world—put more concisely, a strategic concept or orientation—has value in defining broad national goals and priorities and providing the political rationale for domestic and allied support. It gives direction to diplo-

---

[6] ". . . the diversified bureaucratic apparatus of modern states is itself a powerful obstacle to the implementation of any comprehensive scheme of grand strategy. [T]he frequent appeals heard in public life for a 'coherent' or 'consistent' national strategy are not merely vacuous but actually misleading." Edward N. Luttwak, *Strategy: The Logic of War and Peace*, Cambridge, Mass.: Harvard University Press, 1987, pp. 233–234.

[7] David M. Edelstein and Ronald R. Krebs, "Delusions of Grand Strategy," *Foreign Affairs*, December 2015.

[8] Brands, 2014, pp. 14–15.

[9] Ashton Carter, U.S. Secretary of Defense, *Submitted Statement—Senate Armed Services Committee (Budget Request)*, Washington, D.C., March 3, 2015.

macy and defenses, promotes bureaucratic coordination, and enables more-efficient use of resources. Despite the limitations of strategizing and policy planning in a turbulent world—to paraphrase President Eisenhower—while plans in real-world application may turn out to be worthless, planning is everything.[10]

As Paul Miller noted:

> It is true that the world is growing more complex—which is precisely why the United States needs a more coherent and integrated strategic planning capability.[11]

In short, we conclude that the *process* of strategizing is essential, even if the strategies that result must be adapted or discarded over the course of time.

What Is Strategy, Anyway?

In the argot of foreign policy and defense specialists, there is no more ambiguous and adaptable a concept than "strategy." To say one has a *strategy* implies a plan of action to achieve a specific goal or to solve a specific problem. To say something is "strategic" means it is of exceptional significance, if not of game-changing importance. A "grand strategy" has been defined as a nation's "policy in execution,"[12] "the calculated relationship of means to large ends."[13] *Strategy-making* has been characterized as "the central political art . . . the art of creating power."[14]

---

[10] Cited in Freedman, 2013, pp. 609–610.

[11] Paul D. Miller, "Organizing the National Security Council: I Like Ike's," *Presidential Studies Quarterly*, Vol. 43, No. 3, September 2013, p. 604.

[12] ". . . The role of grand strategy—higher strategy—is to co-ordinate and direct all the resources of a nation, or band of nations, toward the attainment of the political objective of [a] war . . ." B. H. Liddell Hart, *Strategy*, Second Revised Edition, New York: Meridian Books, 1991, p. 322.

[13] John Lewis Gaddis, "What Is Grand Strategy?" Heyden Distinguished Lecture, Duke University, February 26, 2009, p. 7.

[14] Freedman, 2013, pp. 609–610.

Strategies can be seen as a hierarchy of conceptualizations, from the highest level of a national "grand strategy"—"ultimate ends and basic means"—to tactical-level actions.[15] Developing effective strategies for dealing with competitors or adversaries requires identifying critical asymmetries, "pivot points," or sources of leverage that can magnify a nation's power relative to its adversaries.[16] Implementation of a strategy may include the use of "stratagems"—schemes or tricks designed to deceive or outwit an adversary. Few uses of the concept of a strategy get down to the precision and detail of a policymaker's operational application:

> A real strategy has to have a tangible plan, it has to have a clear program to implement that plan, and it has to have the budget and resources to make it work. This means making difficult trade-offs and setting clear priorities. It means establishing accountability and having measures of effectiveness. It also means justifying the choices with a clear analysis of the risks and costs involved.[17]

To make the subject of strategy still more complicated, each U.S. administration publishes one or more documents entitled "National Security Strategy of the United States." However, a comparative assessment of the most recent set of these documents reveals that they are not really statements of "strategy." They are aspirational or visionary expressions of the kind of world the incumbent wants to shape, and they list priorities among the many challenges facing the country. They have been criticized as "lists of eminently desirable goals with hardly a hint as to how they might be achieved under existing resource constraints and in the face of active opposition from American adversaries."[18]

The lead volume of the Strategic Rethink project laid out choices for the United States—key issues that will face the administration that

---

[15] Luttwak, 1987, p. 70.

[16] Richard P. Rumelt, *Good Strategy, Bad Strategy*, New York: Crown Publishing Group, 2011.

[17] Anthony H. Cordesman, "Strategy, Grand Strategy, and the Emperor's New Clothes," Commentary, Washington, D.C.: Center for Strategic and International Studies, March 17, 2015.

[18] Andrew F. Krepinevich and Barry D. Watts, *Regaining Strategic Competence*, Washington, D.C.: Center for Strategic and Budgetary Assessments, 2009, p. 11.

will assume office in early 2017. In this concluding overview, we conceptualize three "grand" strategic orientations that reflect alternative views of how to advance U.S. national interests in the turbulent world described in Part I.

Defining National Interests

To begin, however, we should define the core U.S. interests that are—or should be—common to any future administration.

The primary national interest of the United States is to advance the security, freedom, and prosperity of the American people, so that they may enjoy, as the Founding Fathers put it, "life, liberty, and the pursuit of happiness." Over time, national leaders have enunciated a large number of subsidiary interests, sub-subsidiary interests, and doctrines for how to achieve them. These interests evolve over time.

The schema laid out in List 6.1 is intended to help nonstrategists think about essential interests versus evolving ones. The former are narrowly construed because true "interests" (which must be defended) should not be confused with "objectives" (which evolve over time, and will be pursued by most administrations wherever possible but are not essential to the survival of the country). On closer inspection, many so-called "interests" turn out to be mutable. Defending South Korea from attack from the North was not identified as an interest until President Truman decided to intervene in 1950 in the aftermath of the North Korean attack on the south. Access to Middle East oil supplies, however necessary to the Western economies, was not a formal U.S. interest until President Carter declared it to be so. Preventing genocide, however morally commendable, was not deemed a U.S. interest until President Obama announced it would be. Moreover, U.S. presidents sometimes choose not to defend interests that the United States has previously declared, when doing so is deemed too costly or dangerous. Presidents Eisenhower and Johnson did not intervene in Soviet crackdowns in Hungary in 1956 and Czechoslovakia in 1968, although U.S. foreign policy was dedicated to the liberation of the "captive peoples" living under Communist rule. President Obama did not bomb Syria

List 6.1
National Interests and Objectives

National interest:

- Advance the security, freedom, and prosperity of the American people.

Core objectives (required to protect the national interest):

- Protect the U.S. population and territory from nuclear attack and other forms of irreversible harm.
- Increase U.S. economic growth and broad-based prosperity and retain U.S. global economic leadership of the liberal, rules-based economic order.
- Prevent hostile domination of Europe, East Asia, and the Middle East.
- Undercut the ideologies of violent Salafi-jihadism and reduce the capabilities of groups espousing it.
- Make good on U.S. commitments to NATO, Asian allies, Israel, and others, while persuading allies to do more for their self-defense.
- Limit the number of nuclear-weapon states. Combat proliferation of nuclear and biological weapons and leakage of sensitive advanced technologies.
- Advocate respect for universal human rights and protection for freedom seeking people.
- Retain control over U.S. borders to prevent illegal migration or terrorist infiltration.
- Reduce the risk of climate change to the United States.

when Assad used chemical weapons against civilians, violating a previously declared "red line." Thus U.S. "interests" are sometimes aspirational, not a forecast of when a President might or might not respond to a particular provocation.

This list of objectives is neither absolute nor objective—no list can be. It reflects values expressed by presidents of both political parties over many decades, as well as current public perceptions of the most important threats or problems.[19] All represent fundamental issues of safety, security, and prosperity, while also recognizing that American values serve to underpin much of the role the United States plays abroad.

[19] For example, in a 2016 Pew survey, more than 50 percent of those surveyed cited as top concerns ISIS, cyberattack, global economic instability, rapid spread of infectious disease, the number of refugees leaving countries such as Iraq and Syria, global climate change, and China's emergence as a world power. Illegal immigration is seen as a major threat by Republicans, while climate change is seen as the leading threat by Democrats. Pew Research Center, 2016-b, pp. 24–27.

There is no single way to satisfy these interests or objectives; thus, in the chapters that follow, we explore different strategic concepts or orientations. The policies chosen to implement them can be fluid, often dictated not by ideological or strategic considerations, but by fast-changing circumstances, diplomatic necessities, financial constraints, and, above all, judgments about what is likely to work.

For example, persuading allies to do more for their own defense will be an objective for any future administration, but the three strategic concepts would differ in how they set out to accomplish this. Under the first and third options, the United States would honor its existing commitments—for example, the recent pledges made to NATO to help improve deterrence along its eastern flank. But it would not increase aid to allies, even if the security situation were to deteriorate, lest it undercut the motivation for the allies to respond to security challenges themselves. Under the second option, the United States would view deterring Putin as a key objective and would do more in Europe even if the NATO allies were unwilling or unable to boost their own defense spending as much as Washington would like. That calculation could change, however, if Russian leadership became less belligerent, or if another security threat were seen as even more pressing. We elaborate more on these differences in the following three alternative strategic concepts:

- "Come Home America:" Domestic renewal and international restraint
- "The Indispensable Nation:" America as promoter of world order
- "Agile America:" Adapt and compete in a changing world.

We identify the core beliefs and assumptions behind each strategic concept, and interrelate broad policy objectives, degrees of international activism, general budgetary implications, constraints, and risks.

While this report focuses primarily on international strategy, it is impossible to disentangle U.S. domestic politics and policies from their effects on U.S. foreign policy, or the world at large. The behavior of the United States is a key factor in the geopolitical equation on a wide range of issues, from global trade to climate change to the regulation of technology or the promotion of innovation. And the United States cannot succeed in its international goals if its domestic policies are not aligned with these objectives. Therefore, domestic policies that are relevant to broader international policy are also discussed, albeit in brief.

Naturally, the success of any strategy depends on whether the underlying assumptions prove correct, and thus, whether the strategy is well timed and appropriate to the specific circumstances in which the United States finds itself. An illustration of the interaction between strategies and outcomes is shown in Table 6.1. The best outcomes occur when the assumptions that led to a particular strategic choice come true. The worst outcomes happen when the world that was envisioned is not the world that is encountered. The problem is analogous to the dilemma of a homeowner trying to decide how much insurance to purchase. If the United States expends a large amount on national defense and security and the world proves more peaceful than expected, it will have overspent on insurance at the expense of domestic priorities. If it slashes spending for security and international affairs, retreats from its

Table 6.1
Strategies and Outcomes

Strategy Chosen	"Come Home America" Assumptions Come to Pass	"Indispensable Nation" Assumptions Come to Pass	"Agile America" Assumptions Come to Pass
"Come Home America" Domestic renewal, international restraint	**Best**	World more threatening to U.S. interests; nation underinsured	Insufficient innovation; others define rules of global trade
"Indispensable Nation" Promoter of world order	World less threatening; drained resources; nation overinsured	**Best**	U.S. indebted; economic growth suffers; U.S. not well positioned for new opportunities
"Agile America" Adapt and compete in a changing world	Drained resources; U.S. actions create tensions where they may not have existed	Others lead or chaos spreads	**Best**

alliances, and then encounters a more threatening world, it will have underspent on insurance and will face additional costs and risks, some of which may be highly consequential.

These strategies are meant to elaborate on plausible choices for U.S. leadership in the framework of the contemporary political debate. We do not endorse or recommend any of them. We assume that the next administration will mix and match policies as it deems fit. Nevertheless, we hope the conceptual frameworks discussed in this report will help policymakers and the public in considering these difficult choices.

CHAPTER SEVEN

Option I: "Come Home America": Domestic Renewal, International Restraint

> . . . In the new world disorder, America needs national security policies that begin and end by asking what's in these policies for Americans, not what foreign nations long dependent on our protection might think about them. There is no reason for us to continue to shoulder burdens others can now bear. We should build our strength while holding it in reserve. We should act only when it's in our interest to act.
>
> Ambassador Chas W. Freeman, Jr. [1]

The United States has been fighting Salafi-jihadist terrorists for nearly 15 years—the longest foreign war in U.S. history and an inconclusive one.[2] A restrained conception of America's role in the world begins with the premise that both economic vitality and political cohesion, which have underwritten U.S. involvement and influence in world affairs for decades, have been stretched to the breaking point.[3] The view that the

[1] Chas W. Freeman, Jr., "America in the New World Disorder," remarks at the Watson Institute for International and Public Affairs, Brown University, February 11, 2016.

[2] The United States fought repeated skirmishes and wars with various Native American tribes over a much longer period, however. Smithsonian National Museum of American History, "Western Indian Wars," web page, undated.

[3] Foreign affairs columnist Ian Bremmer, who termed his version of this posture "Independent America," argued that the United States must "declare independence from the need to solve other people's problems and . . . finally realize our country's enormous untapped potential by focusing our attentions at home." Ian Bremmer, *Superpower: Three Choices for America's Role in the World*, New York: Portfolio (Penguin Group), May 2015, p. 6.

United States must not to go "chasing monsters"[4] abroad when there are so many unresolved problems at home is as old as the nation itself. It need not mean isolationism. Nor does it imply that a U.S. decline is imminent or inevitable. It may be the price of national renewal.

This strategic orientation would accept that some—indeed many—international problems are beyond America's ability to control. Therefore, it would focus on what is within U.S. control, prioritizing domestic growth and improving U.S. international competitiveness, while attempting to limit military and political commitments abroad. (See List 7.1 for details.) It would reject "liberal hegemony"[5] but would stay engaged in the world for the purpose of advancing U.S. economic interests abroad. Instead of military "boots on the ground," it would prefer to put American-made boots on store shelves around the world. It assumes that both the national security and discretionary domestic budgets will continue to be constrained by debt, slow growth, and political polarization. It would narrow the definition of "core" American national security interests, aiming to *spend* less and *do* less abroad.

In this worldview, a superpower should not have to resort to force so often. The national interest lies in protecting the American people and territory and forestalling war. U.S. diplomacy and other tools of "soft power," backed by economic might, are deemed in many cases to be just as effective, if not more effective, than military force in advancing national well-being over the long term. This is particularly true when considering the indirect and opportunity costs of excessive use of military force. In any case, a fiscally fit United States could always muster resources to deal with a serious security threat, whereas an overextended United States might find itself in a weak position when hit by the next strategic surprise—and there will be surprises. The United

---

[4] Secretary of State John Quincy Adams said, "America does not go abroad in search of monsters to destroy. She is the well-wisher to freedom and independence of all. She is the champion and vindicator only of her own," John Quincy Adams, speech on U.S. foreign policy, July 4, 1812.

[5] Barry Posen used this term to describe the school of thought that led to U.S. overreach. Barry R. Posen, *Restraint: A New Foundation for U.S. Grand Strategy*, Ithaca, N.Y.: Cornell University Press, May 2014.

List 7.1
"Come Home America" at a Glance

- Since the end of the Cold War, U.S. GDP and the well-being of the American people have not increased commensurate with the resources spent on national security.
- U.S. national security will continue to be constrained by debt, slow growth, and political polarization. However, the U.S. nuclear arsenal makes it highly unlikely that any country would attack it directly. The United States can always summon the resources it needs to address truly serious threats. Meanwhile, the money saved should be invested at home.
- The United States will sustain current and promised commitments to NATO and Asian treaty allies, but will do more only when allies do more. It will not automatically intervene to defend the sovereignty and territory of non-allies, unless other vital interests are at stake.
- Russia is highly unlikely to attack a NATO country—and if it did, the aggression could eventually be turned back by a combination of political, economic, and military pressures. China will rise. The primary U.S. interest is avoiding making it an adversary.
- The Middle East may remain in chaos for decades, but the conflicts will eventually burn themselves out. A new political and security equilibrium will be established in the region, but the United States has very limited ability to shape that outcome. Jihadist terrorism is a political problem that will require Muslim leadership; containment may or may not prove possible. A goal for the United States is to avoid another Middle East quagmire.
- America's most important allies are largely able to defend themselves and must do more. The United States will continue to make its military technology available to key allies and offer assistance where needed, but allies must have primary responsibility for their own self-defense.
- The United States can support but not "spread" democracy; democratic change comes from within.
- Globalization is not a goal; expanding U.S. prosperity is. The United States would prioritize "fair trade" over "free trade," supporting only trade deals that expand well-paying U.S. jobs.
- The United States would not embrace a leadership role in reducing dependency on carbon-based fuels. Climate change cannot be countered unless China and India each curtail their emissions, and the United States has no power to compel them to do so. The U.S. government must focus on adaptation to the inevitable impacts of climate change.

States might get its way less often on global security issues, but it could afford to invest in national preparedness and resilience.

The renewal/restraint strategic concept would define a narrower set of U.S. policy goals. As David A. Shlapak put it, "One of the responsibilities of strategy is to place limits on ambition, which means above all seeking to distinguish the things that one cannot tolerate from those one would prefer not to happen."[6]

But which outcomes are intolerable? Under this strategic concept, the United States would defend its treaty allies in Europe and Asia but would not likely intervene militarily to defend non-allies. It would, however, fight to preclude hostile domination of an entire region. Although it might not succeed, it would try to reduce international expectations that the U.S. military will, in the end, always be available to rescue the world from true crises. Allies would need to step up to prevent problems from turning into crises and inviting unwanted interventions by others, as happened in Syria.

This orientation would accept that some—indeed, many—problems are beyond America's ability to solve. Such a worldview poses deep political risk in a nation in which many citizens define American exceptionalism—and thus its moral and political leadership—as the willingness to tackle the tough problems that other nations cannot or will not. Future presidents would need to mobilize both public and elite opinion in support of the position that avoiding quagmires—particularly in the Middle East—are a means to enhance U.S. strength, not a show of weakness. The essential U.S. goal is to protect Americans at home and husband resources to focus on security and economic challenges of higher relevance to the national interest.

Global Economy

Assumptions

All three strategies assume that the United States will continue to have an interest in being the world's largest economy. All would seek to preserve the advantages the United States enjoys, including the primacy of the dollar as the currency of the global financial system. This requires maintaining, expanding, and, where necessary, reforming the rules-

[6] David A. Shlapak, "Toward a More Modest American Strategy," *Survival: Global Politics and Strategy*, Vol. 57, No. 2, March 20, 2015, p. 65.

based economic system that has been the foundation of global growth and prosperity, as well as generating steady domestic growth sufficient to improve domestic standards of living.

The essential goal of the "Come Home America" posture is to renew the sources of domestic strength as the primary means of preserving the U.S. ability to shape and stabilize the world. It differs from the others in two main respects. First, it assumes that mounting debt and political strife have already restricted the country's ability to plan, execute, and sustain an economic or geopolitical strategy. Second, it questions the wisdom of unchecked globalization. It might adopt national policies to boost manufacturing, emphasize fair trade over free trade, and reject international agreements that mandate changes to U.S. regulations. It would, in essence, trade some foreign assistance and the expense of U.S. military power for a higher standard of living at home.

As Shatz argues in an earlier volume of this Strategic Rethink series, three trend lines suggest that many Americans will have trouble achieving rising standards of living.[7] Two of these trends are domestic: the federal budget and the labor market. One is international: weakness in the economies of the most important U.S. allies, the EU, and Japan. This strategy assumes there is little Washington can do directly to improve its friends' economies. It can help indirectly by avoiding war—by deterring Russia and China, working to end the conflicts in Syria and Iraq as soon as possible, and allowing the repatriation of refugees now in Europe. Therefore, the United States must prioritize improving its long-term fiscal outlook and the domestic labor market.

Under this strategic concept, the United States would not seek new international trade agreements and might attempt to renegotiate NAFTA or other existing agreements or to pair them with more forceful measures to combat both job losses and declines in wages for less-educated Americans. It would scrap the TPP and accept that China is likely to be part of whatever agreement follows it. There is some evidence that this could result in more middle-skill jobs. However, the actual effect on the number of jobs is uncertain because there would be

---

[7] Shatz, 2016, pp. 9–25.

job losses from reducing international exchange. Effects could include higher consumer prices, lower overall compensation, and possibly a slowdown in growth of GDP and higher inflation. These would need to be understood as the price paid by the economy as a whole to produce higher wages for some workers.

Russia

Assumptions

Russia will continue to try to regain its influence in the former Soviet territories, but it is highly unlikely to attack a NATO country—and if it did, the aggression could eventually be turned back by a combination of political, economic, and military pressures.

Putin may or may not be deterred by Western economic sanctions, but he might be constrained by Russia's economic woes, including a shrinking economy due to the collapse in oil prices, and an inflation rate of more than 15 percent.[8]

Indirect aggression sponsored by Moscow in other countries may continue. This may include hybrid or "gray zone" tactics, political attempts to undermine European unity, or sabotage.[9] These stop short of being acts of war, which makes them particularly hard to counter, given the strong NATO interest in avoiding war with Russia. The United States and its allies might respond with similar hybrid tactics and/or attempt to counter them by political and diplomatic means.

Under this concept, preventing Russian control or influence over Ukraine or other non-NATO countries is desirable but not essential

---

[8] The World Bank estimated in September 2015 that Russia's economy would shrink by 3.8 percent in 2015 and 0.6 percent in 2016. In November 2015, it pegged the inflation rate at 15.6 percent. World Bank, *Russia Economic Report 34: Balancing Economic Adjustment and Transformation*, September 30, 2015a; World Bank, *Russian Federation Monthly Economic Developments*, November 11, 2015c.

[9] Sabotage and other "gray zone" aggression have been used by both sides; pro-Ukrainian forces are accused of sabotaging the electrical lines leading to Russian Crimea. Ivan Nechepurenko and Neil MacFarquhar, "As Sabotage Blacks Out Crimea, Tatars Prevent Repairs," *New York Times*, November 23, 2015.

to U.S. well-being. After all, the United States flourished for 45 years despite Soviet domination of its "near abroad." Instead, the United States might work with the allies to establish Ukraine as a neutral buffer zone between NATO and Russia, one that participates in both European and Russian trading blocks.

This approach is informed by the view of some Russia experts that the fall of Putin could result in the rise of an even more aggressive, nationalist government in Moscow. Russia could pose a short-term threat to Europe but finds itself in demographic and financial decline over the long term. It is having difficulty keeping its million-man military at full strength, and by some estimates, might be able to afford a defense budget of $100 billion to $150 billion in 2020—only about one-eighth of what the NATO members spent in 2015.[10]

The Russian economy fell into recession in 2015, and even if the economy grows in line with official predictions of a recovery to 2.2 percent growth, this would be far below the average annual rates of the last decade.[11] Labor-force participation has plunged and there is substantial evidence that the Russian people are feeling the impact of Western sanctions, state budget-tightening, and higher food prices under the Putin-imposed restrictions on the import of Western foodstuffs. Anti-American sentiment is running so high, and the media and civil society are now so repressed, that there is little reason to believe that democratic reforms or a pro-Western policy will prevail in Russia any time soon. The essential U.S. goals in this strategy are to protect NATO allies and wait Putin out while attempting to maintain ties with Russian civil society. In the long run, a Russia in decline will eventually have to relinquish its gains in the "near abroad" and perhaps in Syria. To succeed in this strategy, Washington would be required to draw a persuasive red line on attacks—conventional, hybrid, or covert—on NATO treaty allies, and may need to sustain international economic sanctions.

---

[10] Clifford Gaddy and Michael O'Hanlon, "Toward a 'Reaganov' Russia: Russian Security Policy After Putin," *Washington Quarterly*, July 2015.

[11] Keith Crane et al., *Russia's Medium-Term Economic Prospects*, Santa Monica, Calif.: RAND Corporation, RR-1468-RC, 2016, p. 54.

China

Assumptions

China will not attack the United States or its interests directly and it will be expensive—if not impossible—to check all of China's growth in the Asia-Pacific. Even if China continues to improve its military capabilities, a war with the United States would be disastrous for both sides, but much worse for China. The essential U.S. goal is to make sure Beijing recognizes this.

Beijing may continue to play a dual game of profiting from the global trading system while developing an alternative approach to international trade and finance based on its institutions of state-based capitalism. It may also continue challenging some international norms, such as freedom of navigation, and engaging in nationalist shows of strength that discomfit its neighbors. Nevertheless, Washington must manage this ambivalent relationship through diplomacy, as it has done in the past, bolstered by an improving U.S. economy and its longtime military alliances.

This strategic concept recognizes that China will rise; the United States should avoid making it a foe. This may demand a more accommodating stance toward its ambitions to regional primacy—while stressing that regional hegemony may not be achieved by force. The United States should attempt to boost ties with India, while recognizing that India has an independent agenda and is not interested in becoming part of the U.S. regional alliance system. Meanwhile, the United States should encourage Chinese caution by continuing to expand trade and offering Beijing increased authority within global trade and financial institutions, commensurate with its status. Continued cooperation on climate change and innovations in energy technology is in the U.S. interest.

Taiwan will remain a sore point. The United States could continue to provide political and military support to Taiwan, but needs to make clear that there are limits to that support, including provocations from Taiwan that might lead to war.

Middle East

Assumptions

The conflicts in the Middle East will eventually burn themselves out and a new political and security equilibrium will be established in the region. This process could take decades, however. There is no evidence, based on past performance, that the final outcome is likely to be materially improved by U.S. military involvement. In fact, it could make things worse, particularly if the United States and Russia are drawn into conflict or the United States inherits the bulk of the obligation of pacifying Syria or stabilizing it after a cease-fire. The scale of destruction in Syria is so vast that eventual reconstruction will require far more time and resources than the United States invested in Iraq, with no more guarantee of success. Strategists would be guided by the "Pottery Barn rule:"[12] "If you break it, you own it." The United States should avoid intervening in countries that are "broken" unless they have marshaled the resources and will for the protracted nation-building exercise to follow.

Under this "Come Home America" strategic orientation, Washington will negotiate with any rival or adversary—Iran, Russia, even Assad—to try to end the Syrian civil war. But it would work equally hard to avoid entrapment in the conflicts that may follow—with whatever jihadist group might succeed al Qaeda and ISIS, between Iran and Saudi Arabia, between the Turks and the Kurds, or among competing factions in Libya or any other quagmires of sectarian violence. Although the United States will continue to fight ISIS, it should avoid "owning" that conflict. In the long run, the Salafi-jihadist ideology of ISIS cannot be defeated by the West; it can only be discredited by Muslim political, religious, and civil society leaders,[13] along with Muslim military force. Putin may indeed end up "owning" Syria, but will eventually come to

---

[12] Thomas Friedman coined this term. Contrary to media reports, former Secretary of State Colin Powell said he did not use it to warn President Bush against invading Iraq. "But what I did say . . . [is that] once you break it, you are going to own it, and we're going to be responsible for 26 million people standing there looking at us. And it's going to suck up a good 40 to 50 percent of the Army for years. And it's going to take all the oxygen out of the political environment." "Ideas and Consequences," *The Atlantic*, October 2007.

[13] Rabasa et al., 2007.

regret it, just as the Soviet leaders eventually found their intervention in Afghanistan too costly. This policy would require a long period of patience—not a traditional American strength—but it would conserve U.S. power to deal with more consequential challenges, particularly the rise of China. The greatest beneficiary of another protracted U.S. or even U.S.-Russian involvement in the Middle East is likely to be Beijing.[14]

Defense and Defense Spending

Assumptions

The United States cannot care more about the security of other nations than they do. If the United States spends less, allies will have to spend more.

The key U.S. allies are largely able to defend themselves and must do more. The United States will continue to make its military technology available to allies and offer assistance where needed, but allies must have primary responsibility for their own self-defense.

Struggling or not, U.S. European allies will need to meet their pledges to spend 2 percent of GDP on self-defense by 2025, if not before. It will be particularly important that Germany and Japan, which spend 1.3 percent and 1.0 percent on defense, respectively, boost their spending.[15] Japan revised its "peace constitution" in 2015, allowing it to take a more proactive posture, and it sharply increased its defense budget. Still, it is spending less than one-third of what China spends on defense.[16] This would immediately relieve the U.S. defense burden and reduce public resentment over "free riding" by European

---

[14] As Obama adviser Derek Chollet warned, "If the U.S. gets involved in another major ground war in the Middle East, it will be harder if not impossible to sustain the rebalance to Asia." Derek Chollet, *The Long Game: How Obama Defied Washington and Redefined America's Role in the World*, New York: Public Affairs, 2016.

[15] Full a full discussion of the "free-rider" problem, see Hans Binnendijk, *Friends, Foes, and Future Directions: U.S. Partnerships in a Turbulent World*, Santa Monica, Calif.: RAND Corporation, RR-1210-RC, 2016, pp. 84–90.

[16] Binnendijk, 2016, p. 107.

and Asian allies who have prospered under an expensive U.S. security umbrella while spending only half of what the United States has invested in NATO and Japan. The United States would continue to meet its commitments to NATO, including the enhanced measures that have been agreed upon since 2014 to deter Russian aggression. It would also meet its commitments to Japan, South Korea, and other Asian allies, but be extremely wary of making new ones.

This could reduce the risks to the United States of being drawn into conflict. High levels of confidence that U.S. forces would come to the rescue may have inadvertently encouraged allies and partners—including those at increased risk of direct attack—to underinvest in defense even as their security environment deteriorated. Under a restraint posture, the United States would need to make clear that as it allows military spending to decline—though it will still outspend all of its allies—it cannot deploy forces rapidly to three theaters simultaneously. As Binnendijk put it, "Now, partners have a military imperative to be more serious about their own security because the U.S. military could become engaged in another theater and be unable to swing its forces quickly enough."[17] In particular, as discussed in the Chapter Five section on deterrence, vulnerable allies must be better prepared to fend off a rapid-strike attack.[18]

Under a restraint strategy, the United States would abide by the spending caps set forth in the Budget Control Act of 2011 as modified in the Bipartisan Budget Agreement of 2015. Under that budget deal, the President and Congress agreed to allow both domestic and military spending to rise 5 percent (from sequester-constrained levels) for two years, putting spending at about 3.3 percent of GDP.[19] Defense spending was allowed to rise by $40 billion for fiscal years 2016 and 2017, with another $118 billion for OCO spending, including for combating terrorism. Ultimately, this budget would result in defense spending falling to the lowest level of GDP since World War II (see Figure 7.1). It would allow defense spending to decline from 4.7 percent of GDP

---

17 Binnendijk, 2016, p. 54.

18 Ochmanek et al., 2015.

19 Office for the Under Secretary of Defense (Comptroller), 2016.

in 2010 to 3.5 percent in 2014 to 2.3 percent of GDP by 2024. Nevertheless, in real terms, the United States would still be outspending its closest competitors combined, not to mention every one of its allies.

To achieve these spending reductions, the United States would abandon the long-standing defense strategy that required the military to be capable of fighting in two theaters simultaneously. It would accept that the United States will not fight more than one war at a time, and adjust its military posture accordingly. The Army would shrink to below its current manpower strength; the Navy and Air Force would decrease more modestly. Modernization programs would slow as the United States would need fewer power-projection capabilities, and readiness would be "tiered" to place more reliance on National Guard and Reserve forces. Special Forces would be increased in light of their utility in counterterrorism missions. As part of a robust defensive stance, this strategy would enhance missile defenses but would not modernize all of the nuclear triad. The sea-based nuclear deterrent

Figure 7.1
U.S. Defense Spending as a Percentage of GDP

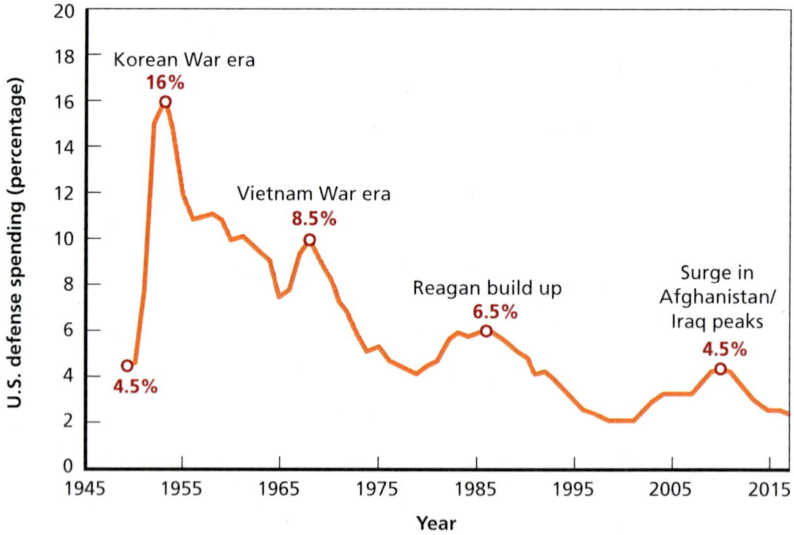

SOURCE: Office for the Under Secretary of Defense (Comptroller), 2016.
RAND *RR1631-7.1*

would be modernized but land-based nuclear capabilities would not. Nuclear bombers would be retained. To the extent they are feasible, arms control agreements would become more desirable under such a strategy, and Washington might consider attempting to negotiate new arms control agreements for nuclear, space, and cyber weapons.

Climate Change

Even if the United States were to meet its own ambitious targets for greenhouse gas emissions, the behavior of other large emitters—notably China and India—will be the determining factor in whether global warming of more than 2° C can be prevented. (China and India have both signed the Paris Agreement but their pledges are voluntary.) If humankind cannot stop the degree of global warming that now seems likely, the U.S. government's duty would be to prepare the country to adapt to the consequences.

Assumptions

Limiting the magnitude of climate change falls squarely within the realm of problems beyond the control of the United States to fix. There are several premises about climate change that would be consistent with a strategy of domestic renewal and international restraint:

- The impacts of climate change over the next decades will prove relatively small, while the cost to the United States and other large economies of attempting to decarbonize quickly will prove unacceptably high.
- The reliance on international norms, inherent in the voluntary pledges of the Paris Agreement, will prove insufficient to induce other countries to cut greenhouse gas emissions.
- China will grow faster than the United States while emitting more.

Whether the Trump administration will withdraw from the Paris Agreement, or attempt to modify it in some way, is unclear. If it chose either to withdraw or modify, other nations might well abandon their commitments, thus ensuring worldwide collapse of the emissions-reduction protocol. The United States would concede international leadership on the issue as too costly and probably futile unless other nations do more.

Even if the United States were to meet the decarbonization pledge it made in Paris, however, and even if the world succeeds in meeting the 2-degree C warming cap—an outcome that looks increasingly unlikely—a domestic adaptation agenda will still be necessary. President-elect Trump has promised a $500 billion investment to rebuild domestic infrastructure, which could pursue a "no regrets" strategy to improve climate resilience. Decisions on where and how to invest in new infrastructure or improvements in water and transportation systems would be scrutinized for their ability to improve efficiency and for national resilience to climate-related damage and natural disasters.

Assuming that climate change will not prove very severe very soon, however, increases the need for a hedging strategy. Investing in the development of cost-effective ways to capture, convert, and store carbon would be one approach. The goals are to store more carbon in trees, plants, and soils, and to deploy technologies to remove carbon dioxide from the atmosphere and sequester it in geologic formations, turn it into rock, or convert it for use in products.[20]

This approach proceeds from the assumption that it may be politically impossible to reduce carbon emissions fast enough to halt climate change, or that it may be cheaper in the future to remove from the atmosphere the excess carbon that will be emitted for some years to come than it will be to reduce emissions now. After all, in the short run, we are unlikely to capture carbon cheaply by adding a filter to every tailpipe of every U.S. vehicle. We could, however, capture carbon from large, stationary sources, including from the coal and natural gas power plants around the country that account for more than 40 per-

---

[20] White House, *United States Mid-Century Strategy for Deep Decarbonization*, November 2016a. A combination of these approaches is known as "bioenergy with carbon capture and storage," in which plants that have absorbed carbon are then burned in power plants that recapture the carbon and sequester it underground.

cent of U.S. carbon dioxide emissions.[21] Therefore, the new President and Congress might increase R&D aimed at reducing the current high cost of carbon capture and storage[22] and increasing the likelihood that the U.S. private sector might develop these technologies first. Of course, even if valuable technologies were developed by a non-U.S. corporation, or by a potential competitor such as China, the United States could nonetheless benefit by buying the technologies at lower cost than developing them.

Constraints

A strategy of "international restraint, domestic renewal" would be most constrained by three factors: U.S. public opinion, fear of terrorist attack, and perceived political risk. Therefore, a strategy of international restraint might entail a flexible, mixed approach. For example, the United States might continue to take a more active military role in fighting ISIS, as this administration has done, while opting for more-accommodating positions vis-à-vis China, supporting more UN peacekeeping, or welcoming allies to lead in smaller regional conflicts, as France did in Mali in 2013.

In contemporary U.S. politics, there are two ideological strains within the "do less with less" camp. Both favor reducing foreign interventions and U.S. defense spending. But some would use whatever funds are saved from buying fewer guns to spend on butter (now known as "nation-building at home"). Others would not spend it at all. They would use any dividends earned from avoiding foreign entanglements to reduce budget deficits. It could be difficult to craft a limited restraint strategy that would win support from both camps, and, moreover, do so without alienating longtime U.S. allies.

---

[21] The Department of Energy estimates that 1,800 to 20,000 billion metric tons of carbon dioxide could be stored underground, the equivalent of 600 to 6,700 years of current emissions. Environmental Protection Agency, "Carbon Dioxide Capture and Sequestration," web page, undated.

[22] Mark Scott, "Companies Struggle to Make Carbon Capture Viable," *New York Times*, October 5, 2015.

Risks

Doing less always runs the risks that adversaries will do more and challenge U.S. preferences. Countries that play a global role bear the risk and financial burden of doing so, and reap the rewards in power and influence. If and when the public mood shifts in favor of activism, the United States could find that power is more difficult to regain than sustain.

The United States would have a reduced capability to project military power around the world at short notice, a capability that is unique but very expensive to build and maintain. U.S. political influence could decline as vulnerable partners would be more easily intimidated by regional powers. Friendly nations that are not U.S. treaty allies, such as Ukraine, might make their own security arrangements and distance themselves from Washington.

The associated reductions in the defense budget would eventually preclude the United States from battling Russia and al Qaeda on two different battlefields, or fighting North Korea while dealing with Iran. DoD spending could be reduced through politically unpopular measures, such as closing of domestic bases, less generous pay and pensions, or cutting weapons systems opposed by the Pentagon, but these would likely raise issues in Congress. However, even cutting such "waste" would not generate sufficient funds to pay for enhanced deterrence, modernizing the nuclear arsenal, or more Special Forces for counterterrorism operations.

In essence, this strategy accepts more military risk in exchange for greater economic might. It would hedge against probable or possible threats but reduce expenditures by not attempting to prepare for less-probable outcomes. After all, the country's track record on intelligence forecasting has historically been poor. The United States will respond to strategic surprises, as it did after Pearl Harbor, the Soviet launch of Sputnik, and 9/11.

The effect of a restraint strategy on the U.S. economy is uncertain. Increased and efficient investments in productive capital could boost growth. The rationale for domestic infrastructure investments is particularly strong, as interest rates are low and there is growing concern among economic thought leaders that forceful measures will be required to boost lagging global growth. For example, IMF Managing Director Christine Lagarde recently warned that the task of stimulating economies has mostly been left to central banks, but monetary policy has been stretched thin, meaning that fiscal policy has a larger role to play. "Where there is fiscal space, record-low interest rates make for an excellent time to boost public investment and upgrade infrastructure," Lagarde argued.[23] The return on infrastructure investments can vary widely, however, and some suggest that remedying inadequate maintenance on existing infrastructure should also be a priority.[24]

Retreating from the world may have the opposite effect, however, by lowering trade and investment.[25] One recent RAND study estimated that a 50-percent reduction in U.S. security commitments would result in a decline of $612 billion, or 0.6 percent of GDP per year for the subsequent five years.[26] It could also raise prices, and the distributional consequences are uncertain. Reducing exposure to trade from low-wage countries and discouraging investment in those coun-

---

[23] Christine Lagarde, "We Need Forceful Policies to Avoid the Low-Growth Trap," IMF Direct blog, International Monetary Fund, September 1, 2016. An earlier IMF paper argued that

> evidence from advanced economies suggests that an increase in public investment that is debt financed could have larger output effects than one that is budget neutral, with both options delivering similar declines in the public-debt-to-GDP ratio. This should not, however, be interpreted as a blanket recommendation for a debt-financed public investment increase in all advanced economies, as adverse market reactions—which might occur in some countries with already-high debt-to-GDP ratios or where returns to infrastructure investment are uncertain—could raise financing costs and further increase debt pressure.

International Monetary Fund, "The Macroeconomic Effects of Public Investment," in *World Economic Outlook, Legacies, Clouds, Uncertainties*, Washington, D.C., October 2014; International Monetary Fund, "Is It Time for an Infrastructure Push?" IMF World Economic Outlook Database, October 2016a.

[24] Nathaniel Popper and Guilbert Gates, "Rebuilding Our Infrastructure," *New York Times*, November 14, 2016.

[25] One widely accepted analysis has found that a 1-percent increase in trade relative to GDP results in per capita GDP growth of between 0.85 percent and 1.97 percent. Frankel and Romer, 1999.

[26] Egel et al., 2016.

tries could boost employment and wages in middle-skill jobs, while at the same time raising prices on goods that lower-income earners buy, thus lowering their standard of living. In addition, employers may have greater incentives to fund labor-saving innovations, so that employment opportunities in some jobs could decline over the medium to long term.

Obviously, this strategy would fail if the leadership of China or Russia proved to be more assertive than expected and willing to risk friction or even conflict with the United States or U.S. partners. It would also fail if it did not improve domestic fiscal stability, economic growth, and living standards for the majority of Americans, to compensate for any decline in its military advantage. Nevertheless, the United States would remain the strongest military force in the world. It would retain its standing as *one* of the leading world powers, although no longer *the* leading power.

The American people would need to accept that other nations—notably China—will seek to expand their influence wherever U.S. power appears to be receding. For example, the TPP agreement does not include either China or India, whereas the Regional Comprehensive Economic Partnership (RCEP), which remains under negotiation, includes all of the ASEAN nations as well as China, India, Japan, South Korea, Australia, and New Zealand, but excludes the United States. China recently announced it would seek accelerated conclusion of the RCEP, in a move to fill the vacuum created by the expected U.S. withdrawal from the TPP, a plan that U.S. partners in Asia could well embrace.[27] Likewise, China has indicated that it would stick with the Paris Agreement even if the United States pulls out, a move that could propel Beijing into a global leadership role on carbon reduction issues.[28]

As discussed, the climate change risks associated with this strategy depend on two currently unknowable facts: the magnitude of climate change on the United States and the rest of the world, and the

[27] "China Wants Quick Close on Regional Trade Agreement for Asia," Bloomberg News, November 24, 2016.

[28] Jill Baker, "Climate Expert Says China Committed to Battling Climate Change—Even If the U.S. Drops out of Paris Accord," Forbes.com, November 14, 2016.

degree of difficulty and cost associated with decarbonizing the global economy. If the impacts of climate change turn out to be large, a U.S. strategy focused solely on adaptation carries significant and potentially irreversible risks for the homeland.

CHAPTER EIGHT

Option II: "The Indispensable Nation": America as Promoter of World Order

> A reconstruction of the international system is the ultimate challenge to statesmanship in our time. The penalty for failing . . . [will likely be] an evolution into spheres of influence identified with particular domestic structures and forms of governance . . .
>
> Henry A. Kissinger[1]

In this worldview, the world without U.S. leadership is a Hobbesian place. Active U.S. leadership and engagement, whatever its flaws, is deemed vastly preferable to chaos. The steeper costs are outweighed by the gains for U.S. citizens and the world. The devastating civil war in Syria, the rise of ISIS, the migration crisis, Russia's military involvement in Syria, and rising instability across the Middle East in the wake of the failed "Arab Spring" are the second- and third-order consequences of premature U.S. disengagement from Iraq and Libya. With all its warts, only the United States has the resources, prestige, and expertise to lead, and the United States is at its best when advancing and defending a benign liberal international world. Failure to invest in global security in this moment of turbulence and change will diminish U.S. power, security, and prosperity—and no other power will shape the world to such benign ends.

Therefore, this strategy begins with the premise that the United States must commit to global leadership and pay for the security it

---

1 Henry Kissinger, *World Order*, New York: Penguin Press, 2014, p. 371.

needs. It would require lifting budgetary spending caps in order to "do more with more." Deficit and debt management are important, but these must be dealt with by long-term measures. Austerity measures applied to U.S. foreign and defense policy would be shortsighted, particularly at a moment when potential adversaries are investing in advanced military capabilities. Resources would be generated, above all, by trying to grow the U.S. economy but also by reallocating spending within or between federal agencies, cutting domestic programs, adjusting entitlements, raising taxes, or accepting more debt. For long-term fiscal stability, both revenue increases and spending cuts would be unavoidable.[2]

As a promoter of world order, only the United States has sufficient means and alliances to deter aggression and promote peace and prosperity; therefore, it should and it must. The more successful the United States is in deterring threats, the more time the next administration would have to work on strengthening and expanding international institutions, partnerships and alliances, humanitarian efforts, global stewardship, genocide prevention, and other global goods.

From a position of fiscal probity and military strength, the United States would be better positioned to conduct the values-based foreign policy that is a hallmark of this strategy. It could speak out for human rights. It would seek public support for the notion that respect for human dignity, acceptance of diversity, and embrace of pluralism are not "Western" values being imposed on developing nations; they are universal values and the foundation for peace and security in a populous, interconnected world. Such a U.S. stance will irritate some partners, as well as adversaries. However, in this view, the United States is always strongest when its values and interests align; a world where the United States does not represent a beacon for freedom and human dignity is a world most of its citizens would not want. Contrary to the concept of national interests in the "Come Home America" option, this strategic orientation defines assistance to freedom-seeking peoples around the world as an important U.S. interest. Moreover, the United States must continue to be the "indispensable" leader of global efforts

---

2 Shatz, 2016.

to combat poverty, human trafficking, violent extremism, and gender-based violence, and to conduct urgent humanitarian operations.

A summary of the key objectives of this option is found in List 8.1. The "Indispensable Nation" option would seek to advance free trade and commit more resources to evidence-based global development efforts, as well as to programs abroad promoting the rule of law, education, entrepreneurship, public health, civil society, and expanding the foundations of democracy.

Supporting these goals would entail increases in the foreign affairs budget. This would support an assertive diplomatic posture. It would improve the security of U.S. diplomatic missions and consul-

List 8.1
"Indispensable Nation" at a Glance

- The United States remains the largest economy in the world, and retains the capability, experience, and expertise to lead. Its political leadership unlocks the resources and builds public support to underwrite robust U.S. international engagement.
- Reversing any perceptions of American weakness or disengagement is a priority, lest adversaries be tempted to challenge U.S. or allied interests. U.S. influence derives from its preeminent security position and ability to support the security of allies and partners. This undergirds stronger U.S. diplomacy.
- Russian and Chinese military capabilities will continue to improve and both countries will engage in risk-taking behaviors to assert regional influence. Investing in deterrence is safer and thriftier than risking conflict with nuclear-armed rivals. Increases in military spending and the foreign affairs budget are both unavoidable and revenues must be raised to pay for the security that the United States needs.
- To maintain the liberal international order, the United States must be prepared to defend the principle that territory cannot be seized by force,[a] although it might not use military force in every instance.
- Key U.S. allies and partners in the Middle East are becoming more fragile. Iran, abetted by Russia and China, could become the dominant power in the Persian Gulf, an outcome that should be precluded.
- The EU is a key U.S. economic and security partner. If it is weakened by the departure of the United Kingdom or other countries, the United States will have a larger role to play in building coalitions among EU and non-EU allies.
- The United States would aspire to be the world leader on reducing greenhouse gas emissions, encourage the development of strong international institutions that encourage all countries to reduce emissions, and provide assistance to the developing world to reduce the most consequential impacts of climate change that will occur.

[a] Dobbins et al., 2015, p. 97.

ates around the world, enabling diplomats and development officials to remain at dangerous but important postings; enhance their ability to move about in difficult environments; upgrade security for civilian and military communications; and, above all, funding for U.S. assistance to key allies and partners. At the same time, the United States would invest in boosting the credibility of its military deterrent, demonstrating its commitment to allies while requiring that they also do more to address deteriorating global security conditions.

Global Economy

Assumptions

U.S. economic engagement abroad promotes U.S. incomes and jobs. The United States would step up negotiation of regional and bilateral free trade accords, vigorously promote U.S. exports, push for congressional approval of the TPP, conclude negotiations for the Transatlantic Trade and Investment Partnership (TTIP), and secure congressional approval for that pact as well. It would not only maintain existing security commitments but might also enter into new ones, in light of the evidence that these alliances and partnerships not only contribute to security but also have significant positive effects on U.S. bilateral trade by increasing both imports and exports.[3]

Russia

Assumptions

Putin's Russia is a threat to international order. The hopes that Russia would join the larger zone of democratic peace have been dashed for now. President Putin is playing a particularly aggressive, dangerous game in trying to reinstall a buffer along Russia's periphery and using military force to do so. Given the state of Russia's economy, to say nothing of its demography, it is possible that Russia may have found

---

[3] Egel, 2016, p. 47.

the limits of its reach. But there is abundant danger in Russia's ongoing probing, and it remains one of the few powers capable of threatening the United States directly.

The United States has an interest in preventing Putin from breaking European and NATO cohesion; it would maintain unilateral sanctions on Russia until Putin pulls out of eastern Ukraine and returns it and Crimea to Ukrainian control. A key challenge will be maintaining and enforcing the international sanctions in the face of weak European economies.[4]

To deter and defend against further Russian aggression, NATO has approved the deployment of four multinational battalions rotating through Estonia, Latvia, Lithuania, and Poland.[5] Depending on Russian behavior, the United States and NATO might seek to enhance this force. While this could be costly, the United States and NATO would pay an even higher price for failure to do so. They would arm Georgia and Ukraine and help them develop effective territorial defense, beginning the process of helping them prepare for eventual NATO membership. Meanwhile, the United States has an interest in protecting the principle of national sovereignty and might fight to defend a non-treaty ally to enforce it.

In this more active role, the United States would counter Russian efforts to dominate the Eastern European media markets with pro-Moscow propaganda. The United States would maintain generous asylum policies for Russian journalists, opposition figures, and civil society activists and would support efforts to disseminate their views, and might consider reestablishing the U.S. Information Agency. The United States would strive to offer a more attractive model than that offered by Russia, China, Iran, or other nations that promise citizens stability and prosperity in exchange for obedience to a nationalist, authoritarian state.

---

[4] William Courtney and Michael Haltzel, "Putin is Trying to Bully Europe into Dropping Sanctions," RAND blog, June 11, 2016.

[5] NATO, "NATO Leaders Bolster Collective Deterrence and Defence," press release, July 8, 2016.

China

Assumptions

China will become more powerful economically and more assertive and less risk-averse in the security domain. Managing the inevitable U.S.-China competition, with associated ups and downs, while maintaining a viable and coherent security framework in East Asia will be a first-order question for the next President and his successors. This option would aim to anchor China in the liberal international political and economic order while finding ways to place appropriate checks on the exercise of its growing military power. The notion of an Asian century dominated by China is not in the interest of the United States or its allies.[6] Nor is a China that continues to view itself as a historic victim of Western imperialism and will thus feel aggrieved against if its aggressive nationalism is checked.

Under this posture, the United States would respond to Chinese naval aggression in the South China Sea that might threaten freedom of navigation in international shipping lanes. It would promote development of stronger Asian security institutions. While making clear that it has no intention of encircling or containing China, the United States would also announce its intention to honor its security commitment to Taiwan. As the promoter of world order, the United States would need to be prepared to defend the principle that territory cannot be seized by force.[7]

The United States would cooperate with China as a major economic partner. It would invite China to join the TPP if Beijing is willing to accept its terms, including investor protection, reduced trade barriers, labor and environmental protection, and protection of intellectual property. Washington would not welcome or promote the internationalization of the renminbi. However, it would work with Beijing to lessen risk of potential financial instabilities or bank failures that could

---

[6] Kishore Mahbubani, *The New Asian Hemisphere: The Irresistible Shift of Global Power to the East*, New York: Public Affairs, 2009; Robert D. Blackwill and Ashley J. Tellis, *Revising U.S. Grand Strategy Toward China*, Washington, D.C.: Council on Foreign Relations, March 2015.

[7] Dobbins et al., 2015, p. 97.

disrupt global markets. The President would assure China of America's desire to harmonize the two countries' positions wherever possible in such international bodies as the IMF and the United Nations.

Middle East

Assumptions

U.S. allies in the Middle East lack the capability to degrade ISIS and offset Iranian power without U.S. assistance. The question is not whether to intervene, but how extensively. A minimum goal will be to see non-jihadist forces retake territory controlled by ISIS, recognizing that conflict among these forces may worsen if they succeed in dislodging their common enemy. It would go on to restore Iraqi control inside its borders, and halt the refugee flow from Syria. Russia and Iran will have a role in any peace talks, but they must cooperate in finding a political solution that will stabilize Syria and represent its ethnoreligious diversity.

A more assertive option is to increase U.S. ground and Special Operations forces and add ground troops to the coalition fighting ISIS, partly to balance Iranian influence in Iraq and Russian involvement in Syria, and partly to help forestall a resurgence of conflict among Shiites, Sunnis, Alawites, Turks, and Kurds once their common enemy, ISIS, is defeated.

The United States will need to decide whether it can accept a stabilized Syria with any government that can keep the peace—probably led by Assad—or whether it is willing to commit substantial military forces and postconflict resources to building a different sort of nation-state—an effort that could engender major political opposition. The United States will keep forces in Iraq for some time to train a more robust army and prevent the breakaway of the Sunni areas of northern Iraq.

Defense and Defense Spending

Assumptions

In light of serious threats now posed by Russia, China, Iran, and North Korea, the United States must reverse the decline in military readiness and deterrent power. It would maintain a robust two-war capability, ensuring modernization efforts are outpacing those of Russia and China. It would increase U.S. ground force strength and bolster its forward deterrence posture. It would invest in A2/AD capabilities at moderately larger levels while helping key allies develop their own protection, and would increase R&D spending, including on nonkinetic tools of deterrence and war.[8] It would develop robust offensive and defensive capabilities in cyberspace and would seek to improve anticipation and build national resilience against a range of possible disasters, attacks, and disruptions.

While we have not calculated the full costs of the activist "Indispensable Nation" option, we estimate that retaining traditional levels of U.S. military preparedness in the face of determined adversaries will cost roughly 4 percent of GDP. Even as the country is spending more, it would want to spend wisely by insisting on a rethinking of new, more cost-effective approaches to deterrence and defense—as envisioned in the DoD "third offset" initiative,[9] which aims to strengthen conventional deterrence but also address hybrid problems, including with advanced technologies that will pair humans with machines.[10]

---

[8] Ochmanek et al., 2015.

[9] In the "first offset," U.S. military planners in the 1950s and '60s built a robust nuclear arsenal to "offset" the superiority of the Warsaw Pact nations in conventional armed forces. The "second offset" in the 1970s countered a Warsaw Pact troop buildup on the German border with advanced weapons based on digital microelectronics. The current "third offset" aims at countering proliferation of advanced weaponry around the world and maintaining U.S. military superiority. Bob Work, Deputy Secretary of Defense, speech at National Defense University, Washington, D.C., August 5, 2014.

[10] Bob Work, Deputy Secretary of Defense, speech at the Reagan Presidential Library, Simi Valley, Calif., November 7, 2015.

Climate Change

Assumptions

Climate change poses a serious threat to stability and prosperity and it is in the U.S. national interest to lead the world in aggressively pursuing both adaptation and mitigation at home and abroad. Proceeding from this assumption, it might pursue policies to incentivize rapid climate adaptation, accelerate R&D aimed at developing zero-carbon technologies,[11] and/or begin diplomatic initiatives aimed at introducing a unified global carbon-pricing mechanism that would incentivize decarbonization by private-sector companies all over the world and defray the costs of adaptation and mitigation.

Further, this strategy would be compatible with the following policies:

- increasing federal R&D spending on clean-energy innovation and leveraging global public-private partnerships to speed the effort
- pushing for a rules change in the WTO to protect the competetiveness of countries that tax carbon emissions[12]
- imposing a carbon tax, with offsetting tax credits or other mechanisms that would lessen the burden on low-income Americans and/or low-emitting businesses and provide incentives for the decarbonization of the U.S. economy[13]
- meeting or exceeding the greenhouse gas emissions reduction targets that were declared at the December 2015 Paris summit and agreed on by the United States and China, and assist other countries in doing so

---

[11] This is the preferred approach of the "Breakthrough Energy Coalition," a group of billionaires who have pledged $100 billion to fight climate change, arguing that joint efforts by private industry, government, and philanthropy will be required to create affordable clean-energy solutions.

[12] William D. Nordhaus, "A New Solution: The Climate Club," *New York Review of Books*, June 4, 2015.

[13] Such a tax could be income-neutral or designed to generate revenue for other purposes. It also could be the basis for urging other high-emitting countries to adopt similar approaches.

- reorienting development aid to prioritize climate adaptation and mitigation.

Constraints

Absent a strategic shock or mass-casualty terrorist attack on the homeland, U.S. leaders may find it difficult to unlock the significant resources required for success in this posture. They would need to present a convincing rationale to garner public support for taxes or borrowing to support another round of Middle East stabilization, if not nation-building. The U.S. public is far more likely to support strengthening military capabilities to counter Russia and China.

Even were President-elect Trump to decide not to withdraw from the Paris Agreement on climate change, the investments needed to support a rapid, economy-wide push for a zero-carbon technology revolution would likely run into substantial competition with higher priorities, such as infrastructure redevelopment or national security. A carbon tax sufficient to produce rapid changes in private-sector behavior would likely be constrained by his own party as well as the Democratic opposition, unless it were incorporated as an element of a package that would reduce and reform overall U.S. corporate tax rates.

Money is only one of the hurdles for public acceptance of an activist foreign policy. A deeper issue is the public loss of faith in the U.S. ability to succeed in stabilizing the chaos in the Middle East or other troubled areas of the world. As David Brooks put it, "It's not clear the foreign policy and defense apparatus believes anymore in its own abilities to establish order, or that the American public has any confidence in U.S. effectiveness as a global actor."[14] Yet restoring public faith in the government's ability to "win," or prevail in a military or political struggle, is becoming increasingly difficult in this multipolar world.

An activist strategy is also constrained by the U.S. public's willingness to sustain it over the periods of time necessary to succeed at many—if not most—peace-building endeavors. U.S. leaders would

---

[14] David Brooks, "Enter the Age of the Outsiders," *New York Times*, October 20, 2015.

need to make a persuasive case that the dangers of overreach are outweighed by the costs of losing relative power to rivals who do not share its values. Unfortunately, this case is more easily made in the wake of tragedy.

A European Union weakened by the departure of the United Kingdom or other important members will be a less effective partner for the United States. It will consume more U.S. diplomatic resources, create more complexity in trade, and constrain the agility of the U.S.–EU alliance. Washington should do what it can to encourage other EU members to stay.

Risks

Overreach remains the chief risk for the "Indispensable Nation" strategic orientation. Unmanageable chaos in the Middle East, worsening relations with Russia and China, sluggish economic growth, worsening federal debt, and weak allies could all thwart U.S. plans. Specific risks include:

- The Middle East quagmire continues. Conflict persists in Syria, Yemen, Libya, and/or elsewhere, sending more refugees into Europe. U.S. involvement drives more jihadists to the ISIS cause. Or, ISIS is ousted from Syria and Libya only to take root elsewhere, or transform into a new terrorist entity that can wage war on the West even without a territorial base.
- Public support fades as time goes on and U.S. casualties mount.
- U.S. military buildup frightens Russian and/or Chinese leadership and prompts them to take a more aggressive stance if they see Washington as threatening. An arms race may ensue.
- There might be an increased risk of being required to contest areas of minimal U.S. interest with Russia or China in the process of defending allies.
- Tax hikes and/or increases in debt are required. Either might weaken the U.S. economy.
- An activist American international profile increases the nation's exposure to hostility and attack.

- Asserting U.S. leadership on climate change only works if the United States backs its words with actions. Failure to invest in climate action would undermine U.S. credibility. Further, this strategy runs the risk of overcommitting national investments if climate change impacts end up being small (less than 1 percent of GWP per capita at the end of the century), or decarbonizing the world economy proves relatively costly (more than 5 percent of GWP per capita). These risks would be present even if the global consensus process were to fail. In that case, the United States would be exposed to even further risk unless it were also pursuing bottom-up processes simultaneously as a hedge.

CHAPTER NINE

Option III: "Agile America": Adapt and Compete in a Changing World

> The price of doing the same old thing is far higher than the price of change.
>
> Bill Clinton[1]

> Only a donkey never changes his mind.
>
> Israeli proverb[2]

This strategic orientation accepts the inevitability of continued globalization and the likelihood of rapid change in the coming decades. Its goal is to manage these changes in a way that benefits all Americans. In this view, it will be nearly impossible, and probably undesirable, to try to limit U.S. exposure to an unstable world, as retreat would diminish U.S. security, along with economic and political opportunities. At the same time, the cost of remaining "the indispensable nation" will be unaffordable. The United States would remain fully engaged in the world, but with foreign policies prioritizing its economic well-being and limiting what appear to be expensive security commitments, particularly with partners able to do much more on their own behalf. It

---

[1] William J. Clinton, "Remarks on the Economic Program in Santa Monica, California," February 21, 1993.

[2] Famously cited by the late Israeli Foreign Minister Moshe Dayan after he changed a key negotiating position in order to conclude successful peace talks with Egypt.

envisions a retooled approach to globalization to serve the American people amid changing U.S. geo-economic and security interests.

In this concept, policymakers would not merely seek a middle ground between international restraint and activism; rather, they would attempt to reconceive America's role in a world that continues to change and will be substantially altered (see List 9.1). A muddle-through, reactive approach in which U.S. leaders clutch hard to a 20th-century vision of American primacy will not build the economic strength, flexibility, competitiveness, innovation, or sense of shared national purpose that the country needs to prosper over the coming decades. This strategic concept would unshackle the United States from the foreign policy of the past in order to allow the country to pursue a 21st-century agenda of prosperity and opportunity.

Here, policymakers assume that global change will be at least as rapid and disruptive as in the past, and probably more so. Developing and poorly governed countries are particularly likely to experience increased stress and instability for the next several decades at least, stemming from the addition of another billion people to a warming planet. The economic geography of the world will change, shifting dramatically toward Asia, and U.S. interests will follow. The United States may need to build new security infrastructures that align with its new pursuits, and be willing to abandon or at least downgrade old ones that no longer serve. Opportunities will abound in the world, but the United States will be unable to seize them if it is bogged down in defending "interests" whose saliency is fading.

The United States is well equipped to thrive in the intense global competition of the 21st century. As discussed in Chapter Four, its economy is the most vibrant and resilient in the world. Its labor force is highly skilled, educated, and adaptable. Its private-sector institutions know how to drive continuous improvement in all processes relevant to their businesses. This cycle of continuous innovation in the use of technologies or management practices is even more important than the technological innovation itself. The country's energy future is brighter than at any time in half a century. Falling oil prices benefit U.S. consumers, while producers stand to benefit when prices inevitably rebound. The U.S. higher education system—despite its high cost—remains the envy of the world.

List 9.1
"Agile America" at a Glance

- The United States may not dominate the world economy, but it should be the world's most competitive and innovative economy. National security derives from prosperity.
- Remain fully engaged in the world and attempt to expand the liberal economic order, but be more selective about the use of military force. Economic statecraft is a highly underdeveloped element of U.S. foreign policy.
- Globalization will make American alliances, international partnerships, and friendships more valuable than ever. Cultivate them by emphasizing an agenda of common prosperity.
- Global change will be at least as rapid as in the past, and the United States must be more flexible to take advantage of geo-economic shifts. The United States will reassess the saliency of interests and move away from those that no longer serve its current needs. Managing the economic relationship with China is the top priority. Some accommodation of China may be necessary and even advisable, but not at the cost of ceding East Asia to a Chinese sphere of influence.
- Manufacturing jobs will not return to the United States in large numbers. Only the development of a highly skilled, better-educated workforce can support high-wage employment, entrepreneurship, and GDP growth.
- Russia will seek to discredit the United States as a security partner, and divide and weaken NATO. It might well pressure or attack non-NATO members and possibly even vulnerable NATO states. But with proper cooperation among partners, Russia can be dissuaded and deterred.
- U.S. interests in the Middle East are declining over time and so should U.S. involvement. As in the restraint option, radical jihadist ideology cannot be "defeated," certainly not by non-Muslim powers.
- Serious impacts of climate change can still be avoided. The best way to ensure this is to develop new technologies that can reduce the cost of decarbonization. The United States would consider a carbon tax to speed this process. The best way to build an international climate regime is to do so through a "bottom-up" series of bilateral and multilateral agreements.[a] Technology-driven decarbonization will also generate opportunities for U.S. businesses.

[a] D. G. Victor, J. House, and S. Joy, "A Madisonian Approach to Climate Policy," *Science*, Vol. 309, 2005.

And while basic infrastructure is in need of investment, many key urban centers are undergoing a renaissance and rebirth.

Moreover, the United States still remains the partner of choice across most of the globe. Its diplomatic, commercial, and cultural relationships in Europe, East Asia, Africa, and Latin America remain strong and productive. Commerce with Latin America and Africa has expanded. But for the Middle East, global violence is at historically low levels, although growing strategic challenges from Russia and China require careful attention. The Middle East remains the one region in chaos. Even there, the region may no longer be as strategically significant as it once was, and it is possible to construe policies to limit the greatest harm without putting larger U.S. interests and goals at risk.

This strategic concept would put more weight on expanding economic alliances than the other two options. In this worldview, globalization only increases the value of positive U.S. relationships with other countries, in the diplomatic, financial, commercial, and military spheres. Simply put, the United States derives enormous advantage from having more friends, and more-valuable friends, than any of its potential adversaries. This strategic posture would attempt to leverage this strength. To do so, the United States must be more willing to share geopolitical power with its partners—not only to gain in the rewards, but to also distribute the burden of international humanitarian and stabilization missions. And its allies must do more—much more—to promote international peace and security. Without such a combination of burden-sharing and selective realignment, the United States will find itself increasingly drained, financially and militarily, just at the moment when its demographic makeup will necessitate more domestic spending on health care and pensions. Long-term though these trend lines are, the need for action to forestall the worst outcomes is urgent.

U.S. leaders would redefine American exceptionalism to mean an exceptionally creative and adaptive country, one dedicated to promoting common prosperity and shared security. The United States would aim to be first among equals, a posture more suited to a world in which power is diffused among many states and increasingly shared with nonstate actors, global civil society, corporations, and empowered individuals.

This would entail changes to America's view of itself. The success of this strategy, even more than the other two, depends on the ability of U.S. leadership to articulate a new set of goals, if not a national mission, as a way to move past domestic political dysfunction. The United States cannot conduct business as usual and expect better outcomes in an increasingly challenging and competitive world.

Unlike the previous two strategic concepts, which represent the traditional ends of the U.S. pendulum on foreign policy, this approach would shun ideology and aim for a thorough reassessment of U.S. interests. While the United States has benefited enormously from the liberal international order of the past seven decades, this worldview finds it also much constrained by geopolitical calculations made in the last century that no longer serve U.S. interests in this one. It would begin by updating a number of Cold War and post–Cold War assumptions:

- **A nation-state–based system is the foundation of international order, and the United States must lead the liberal democratic nations in opposing tyranny, in the form of Communist or highly authoritarian states.** Westaphalian and 20th-century notions of stability based on nation-states and political blocs of states are less apt to capture the fluid nature of power in the world as it is today and in the world we are likely to encounter. A more flexible approach is in order in light of such trends as the rise of a suprastate, such as the EU; the advent of powerful non-state actors, such as al Qaeda and ISIS; the emergence of global corporations with profits that exceed the GDP of many nations and that are able to shift jobs, supply chains, capital, and taxable profits across continents; the emergence of powerful civil society groups and geopublics that may be able to check or even trump the policies of many national governments; and changes in warfare. The United States is in its 15th year of war against a nonstate actor, the Taliban. In many shattered places—Syria, Iraq, Libya, or Mali—the notion that statehood equals security is archaic. U.S. forces in North Africa are not attempting to "fix" artificial or failed states; rather, they are attempting to forge counterterrorism cooperation with whatever entity exists that can provide some degree of effective governance—tribal chiefs, local militia commanders, town leaders. Statehood is sometimes a fiction, and not always a desirable aspiration.
- **The United States must not only lead the world in industrial and economic power but also shape an international trading and economic system that favors its allies and partners and**

restricts the growth of its adversaries. This overstates and misdefines the nature of American power today, and creates expectations at home and abroad that are unlikely to be met. The United States may stay ahead of China for some time, but it will not dominate a world of dozens of wealthy democracies and developing nations with far higher growth rates and ambitions to match. It recently failed to push its close allies into line when Washington unwisely lobbied them not to join the Chinese-led AIIB. Some loss of primacy is inevitable, but more loss of influence may occur if Washington attempts to overplay a strong but no longer dominant hand. Shifting from "the United States as Number One" to "First Among Equals" will require U.S. leadership to create a domestic political climate that allows for more international compromise and, in turn, expects more from other countries. The United States should continue to try to shape the rules of international trade and finance, keeping them open and fair, but this will require flexibility, openness to change, and the ability to work through fluid coalitions, particularly as other rising powers redefine and assert their interests.

- **Middle East stability is an important U.S. interest and disengagement would lead to a damaging loss of influence, as well as potential chaos.** In this worldview, Middle East stability may be decades away, but the value of Middle East oil to the U.S. economy will diminish more quickly, and the sooner the United States can scale down its regional military presence and share peacekeeping duties in the region, the better. China and India, which are more dependent than the United States on Persian Gulf oil, and Europe, which is most exposed to migration from and through the Middle East, must assume more responsibility for regional stability, under a UN mandate if possible. They will have no incentive to do so unless the United States catalyzes such a shift. The United States need not abandon traditional allies, but it would signal that it must shift resources away from the Middle East and Europe toward Asia or other emerging markets. In this view, the United States cannot contain and must accommodate and try to benefit from China's rise, while competing for markets and relationships in booming areas of East and South Asia, Latin America, and Africa.

- **NATO and the G-7 are indispensable lynchpins of Euro-Atlantic security and prosperity,[3] and the United States must work hard to strengthen and maintain such institutions.** These organizations remain important, but the United States reaps less benefit from each than in the past, and might invest a bit less in maintaining them. The G-7 has been largely supplanted by the G-20 (an international forum for the governments and central bank governors from 20 major economies founded in 1999) in its role of global economic stewardship. G-7 members no longer represent the world's leading economies. From a military standpoint, Russia is not the Soviet Union, and the Europe of 2017 does not resemble the ravaged continent on which NATO was formed in 1949. NATO and the G-7 are useful in maintaining Western cohesion (more so if China becomes a Western adversary), but neither organization includes all of the friendly Western nations or the important, friendly democracies with whom the United States needs to coordinate economic and military policy. The United States should remain in NATO and do its fair share, but the Europeans who benefit most from NATO must now take the lead in deterring Russia and providing for their own security. This is even more true as Britain leaves the EU. Further expansion of NATO would not serve the U.S. interest. In the long term, the United States should seek to build a trilateral alliance with Asian allies and European allies to better address 21st-century U.S. security needs.[4] Meanwhile, the busy U.S.-NATO calendar—a relic of Western diplomacy in the last century and before the Information Age, should certainly be pruned.[5]

---

[3] The Group of Seven (G-7) includes the United States, Canada, France, Germany, Italy, Japan, and the United Kingdom, and meets annually to discuss global economic governance, international security, and energy policy.

[4] Binnendijk, 2016.

[5] The United States now participates in eight major NATO events per year, including ministerial conferences and side meetings of the allies' defense and foreign ministers, but not

- **Overwhelming military superiority is necessary to protect the United States and advance national interests.** Military superiority remains a goal, but increasing the U.S. military presence in many areas of the world will have diminishing returns. The U.S. military posture must be calibrated for a nonpolar world—enough to deter major-power potential adversaries, but not so much as to provoke major opposition. The United States will want the world's best military, yet in the 21st century, military instruments are becoming more difficult to use; economic instruments of coercion are becoming more desirable. New areas of competition, such as space and cyber, will become as important or more important than static measures of ground, air, and naval forces. As globalization advances, economic statecraft will become more central to U.S. foreign policy.[6]

Recognizing that U.S. strength is based on the dynamism of its private sector, the United States would shift fiscal, tax, and regulatory policies to increase productivity and thereby competitiveness. It would reform tax policy to collect revenue more efficiently and put U.S. companies on a more even footing with their foreign competitors. Assuming that the United States must compete with major powers that have national industrial policies, it would increase spending on basic science and R&D efforts in promising sectors where private capital will not invest heavily enough, fast enough, or for long enough to address critical problems.

Mindful of the economic as well as human costs of strategic surprises and natural or man-made disasters, the United States would also focus on increasing resilience at home and extend these efforts abroad in concert with its closest allies.

As in the first "restraint and renewal" strategy, this third "adapt and compete" option also rests on the premise that U.S. national secu-

---

including regular presidential summits. Preparation for these meetings alone consumes a large portion of the diplomatic calendar for senior State and Defense officials.

6 Robert D. Blackwill and Jennifer M. Harris, "The Lost Art of Economic Statecraft," *Foreign Affairs*, Vol. 95, No. 2, March/April 2016.

rity ultimately derives from prosperity. But it would emphasize that U.S. prosperity depends also on the prosperity of its allies and partners. It would rededicate the United States to an agenda of fostering inclusive and sustainable domestic and international economic growth as the basis for peace and prosperity and the justification for a continued U.S. leadership role in the global economic system. It would commit to advancing the TPP and TTIP trade pacts, and might favor opening a new round of global trade talks. And it would animate domestic entrepreneurship, innovation, and R&D in all major economic sectors. To reduce the demand for and use of costly military force, it would continue global development aid at existing or slightly higher levels and would invest in enhancing conflict prevention and counterextremism efforts and disseminating best practices.

Finally, since the test of leadership, by definition, is the ability to attract followers,[7] this strategy would aim to boost the value and attractiveness of the U.S. global "brand" via soft power. The United States might lead in global stewardship, for example, by making meaningful contributions to mitigating climate change, halting pandemics, or peaceful, sustainable development in Africa and/or the Arctic. The United States would continue to be an active participant in international climate agreements.

A refocused agenda would require U.S. leaders to persuade their publics that national well-being is best advanced by a United States that does not squander blood and treasure in an attempt—perhaps futile—to retain overwhelming military predominance. The United States would continue to play a global leadership role, but it would recognize, as Joseph Nye put it, that "leadership is not the same as domination."[8]

---

[7] Dobbins et al., 2015.

[8] Joseph S. Nye, Jr., *Bound to Lead: The Changing Nature of American Power*, New York: Basic Books, 1990, pp. 182–188.

Global Economy

Assumptions

The United States cannot dominate the world economy but it can become the world's most competitive and innovative economy. International crises must not prevent it from investing in its own people and future. A nation with global economic interests cannot afford to sit on the sidelines. The decline of trade and investment barriers worldwide has brought an increase in trade and investment relative to the size of the U.S. economy. This increases the potential for negative and positive effects as a result of greater U.S. exposure to the economies of other major nations. This strategy would emphasize the need to get America's own fiscal house in order while safeguarding, maintaining, and broadening the liberal rules-based global economic system and its institutions. This could have large, positive, cumulative effects on U.S. growth and economic performance. Selective engagement or retrenchment may allow U.S. policymakers to focus temporarily on domestic issues, but that will also allow other countries to reshape the international trading and investing rules in ways that might work against U.S. interests, and accordingly could have large, negative, cumulative effects on economic performance. Because the United States has global economic interests, and because these are tied to its security interests, there is little choice but to engage globally.

While proponents of all three strategic concepts would agree on the need to increase U.S. prosperity, this strategy would emphasize the need to do so through a dual-track approach of pursuing international trade, which benefits the United States overall, while pursuing policies explicitly designed to increase employment and incomes for the bottom third of American families, those who face the largest employment challenges. There are a wide range of policy proposals aimed at accomplishing this: improving STEM education to better equip the workforce to adjust to a new economy where low-skill, well-paying jobs will no longer exist; reducing the regulatory burden on businesses; and increasing federally subsidized job training and retraining, particularly to allow workers who have lost medium-skill, medium-wage jobs as a result of foreign competition to qualify for higher-paying jobs. It might also involve providing assistance to those parts of the population that cannot be readily retrained to participate in the new fields of endeavor. In this regard, some propose to help displaced workers relocate to look for work,[9] increase the earned-income tax credit, or replace it entirely with a basic income stipend and eliminate the plethora of other anti-poverty programs.[10] Other steps could include tax policies to incentivize domestic job creation; corporate tax reform aimed at keeping corporations, their jobs, and their profits on U.S. soil; and more-active diplomatic engagement to remove barriers to exports and protect U.S. patents and other intellectual property from global infringement.

Another step with potentially high payoff would be increasing infrastructure funding. The U.S. rate of spending on infrastructure is the lowest it has been since World War II, and opportunities abound for renovations, repairs, and the installation of new, more-efficient technologies to replace aging ones. With interest rates at exceedingly low levels, the payoff to such investments could be compelling.[11] Improved infrastructure would not only make people safer and healthier, but additional capital could raise labor productivity. Investments in climate-resilient redevelopment could also lower future costs. Aside from public investment, policy reforms to stimulate private investment could also raise labor productivity and wages.

For fiscal and/or ideological reasons, the next administration might attempt to constrain increases in total spending, yet focus discretionary spending on programs aimed at preventing the hollowing-out of U.S. economic power. Specifically, it would attempt to leverage and expand on existing domestic strengths, through a range of policies to promote innovation, science and technology, and education—and, thus, job growth and competitiveness. It would attempt to build on the U.S. comparative advantage in technology, and in a culture that enables rapid commercial and human adaptation.

Implementing policies might include:

---

[9] "Can She Fix It?" *The Economist*, April 23, 2016.

[10] Paula Dwyer, "A Basic Income Should Be the Next Big Thing," *Bloomberg View*, May 2, 2016.

[11] "Larry Summers Reflects," *Harvard Magazine*, April 15, 2016.

- **incentivizing research and development.** As discussed, the federal government could increase support for basic scientific research in areas where evidence suggests that commercial firms will not invest, either because the likely profit margins are too low or because the timeline for return on investment is too long.

  The next administration also might wish to consider supporting a carbon tax or cap-and-trade market system to incentivize low-carbon technology development, and basic energy research. It might also expand research on basic science relevant to medicine and health policy, and on R&D of treatments and devices that advance public health but are unlikely to be commercially viable. This may have long-term benefits, as U.S. health care costs will consume an ever-larger share of GDP as the population ages, absent changes to the cost and efficacy of American medicine. Internationally, medical knowledge and technology present both a commercial growth market and a realm where U.S. interventions have alleviated misery and generated global goodwill.
- **recruiting and retaining global talent, increasing domestic talent supply.** U.S. universities are a beacon for international students, although they are unaffordable for many. The U.S. knowledge economy will benefit most from attracting and training top talent who will stay in the United States. Proposals for implementing policies include increasing the number of H-1B visas for high-skilled technical workers who will remain in the United States; training and supporting talented students in STEM programs who will also remain in the United States for a defined period. This would help redress the projected inadequate supply of American graduates in STEM programs, which was discussed in Chapter Four. This should be matched by increasing financial support and other programs for American students to improve overall U.S. college graduation rates and particularly address the low completion rates of American students in STEM fields, from apprenticeship to doctoral programs. The next administration might consider pilot programs designed to test the effectiveness of these and other proposals for improving the skills and competitiveness of the U.S. workforce.

- **economic diplomacy.** The growth-and-innovation focus envisioned in this strategy would be bolstered by more robust U.S. economic diplomacy around the world.[12] The United States must be second to none in its ability to advocate for U.S. businesses abroad, along with the rule of law, intellectual property rights, and national institutions that support them. An administration pursing this strategy may wish to expand support for the economic activities of U.S. missions abroad and for Commerce, Treasury, and other federal agencies that are increasingly involved in international activities.
- **foreign aid.** The United States uses an increasingly disciplined process to justify its foreign aid, having concluded that giving countries nontransparent slush funds is not a good long-term strategy.[13] It uses Economic Support Funds and foreign military financing to help friendly countries,[14] but these accounts are constrained by budget caps. It is all the more important for the United States to skillfully use its "soft power," the ability to unleash public loans and private capital, the appeal of its huge markets and the strength of its economic model, to win friends and influence and catalyze reform abroad. Evidence that direct foreign aid improves growth in the target country is decidedly mixed, and aid delivery is certainly problematic, but there is broad consensus that aid has helped advance social welfare, especially health and education.[15] Countries have also engaged in fiscal, governance,

---

[12] Blackwill and Harris, 2016.

[13] The China-led AIIB and the New Development Bank, however, need not impose the conditions that Western donors impose on their borrowers, such as transparency, governance reforms, or measures to protect the environment. They could also insulate the founding countries from Western financial sanctions. Shatz, 2016, pp. 101–106.

[14] Foreign military financing is about $5.6 billion per year, while the Economic Support Fund is about $5.1 billion. U.S. Department of State, *Congressional Budget Justification, Department of State, Foreign Operations and Related Programs, Fiscal Year 2015*, Washington, D.C., March 4, 2014.

[15] Tony Addison and Finn Tarp, *Lessons for Japanese Foreign Aid from Research on Aid's Impact*, United Nations University—World Institute for Development Economics Research, WIDER Working Paper 2015/058, August 2015.

and education reforms in order to qualify for aid from the U.S. Millennium Challenge Corporation,[16] suggesting that aid contingent on positive policy steps—in other words, aid as a form of smart power—can be effective.

The U.S. ability to inflict financial pain on its adversaries is substantial. This ability is multiplied when it acts in close coordination with the EU and other partner governments. Unilateral U.S. sanctions are generally seen as harming Americans more than their targets. However, the use of economic coercion as an alternative to the use of military power is growing. Some argue that Russia has already been adequately deterred from further aggression in Ukraine by the adoption of tough, coordinated U.S.–EU sanctions. Others note that sanctions always take a great deal of time to bite, during which time the adversary may have scored large gains. Moreover, overuse of sanctions could incentivize nations to pursue financial arrangements outside the U.S.-led international system. The next President and Congress might expand efforts by the Treasury to devise even smarter sanctions.[17]

Russia

Assumptions

Europe is ultimately responsible for defending itself against Russia militarily, with assistance from the United States, but Europe and the United States must make common cause to counter Putin's assault on international human rights norms and the promotion of authoritarianism.

As in the earlier "restraint" posture, a global growth strategy concept would expect the United States would join forces with NATO and the EU to confront Moscow but expect its European partners,

---

[16] However, it hasn't moved the needle on political rights and civil liberties. Bradley C. Parks and Zachary J. Rice, "Does the 'MCC Effect' Exist? Results from the 2012 MCA Stakeholder Survey," *MCA Monitor*, Center for Global Development, February 2013. For an overview of the aid effectiveness issue, see Shatz, 2016, pp. 112–117.

[17] Juan Zarate, "Treasury's War: The Unleashing of a New Era of Financial Warfare," *Public Affairs*, New York, 2013.

especially Germany, to lead. The goal would be to wait out Putin while focusing on ways to squeeze the Russian Federation budget and the foreign business dealings of the inner circle of oligarchs. Such a strategy would take time to achieve results, but time is on the West's side—as long as EU and U.S. sanctions can be maintained.

China

Assumptions

China cannot be "contained;" the United States must do a better job of competing with it, while maintaining a viable and coherent security framework in East Asia.

The overall assessment of China does not differ under this option from that under the previous "Indispensable Nation" posture. Both would aim to anchor China not only in the global economy but also in the liberal international order, while finding ways to place appropriate checks on the exercise of its growing military power. However, this option would recognize a strong U.S. interest in continuing robust trade with China, the substantial investment that U.S. companies have already made in that market, and China's growing global financial power. It would therefore prioritize fair trade, full access to Chinese markets, the protection of U.S. intellectual property rights, and the development of Chinese rule of law, particularly its judicial system. Rather than engage in a futile attempt to "contain" China, the United States would try to bind it ever more closely into the international economy and trading system, including supporting Chinese plans to internationalize the renminbi.

China's rise to dominance seems less inevitable today than it did even a few short years ago. Indeed, our late colleague Charles Wolf suggested that the right question to pose is not when but whether China will surpass the United States in GDP.

Manufacturing jobs in China may move to countries with lower labor costs, including U.S. partners. Already, Foxconn, which makes the iPhone, is replacing 60,000 Chinese workers with robots, prompting suggestions that the competition of the future will be between

American robots and Chinese robots.[18] Meanwhile, state-owned enterprises could prove a major drag on China's economy. Washington might accommodate some Chinese interests in the South China Sea, provided they were not attained by force, but it would not compromise on the basic sovereignty of allies or the principle of freedom of navigation. The United States cannot pursue new economic opportunities in Asia without safe sea lanes. DoD would be instructed to follow a slower but steady schedule of upgrading its capabilities (as opposed to a policy of rapid readiness in the Indispensable Nation option).

Middle East

Assumptions

The United States should not allow the Middle East to dominate its agenda, particularly at the cost of neglecting its interests in Asia and other emerging markets. Regional hegemony, even if it could be achieved, is not worth the price in U.S. blood, treasure, and opportunities lost.

The United States has not received the hoped-for return from nearly four decades of intensive engagement in the Middle East. Although President Obama was not able to conclude the wars in Afghanistan and Iraq and draw down U.S. forces, that should remain the goal. The United States may not be able to revert to the role of "off-shore balancer" as there may be no balance to maintain. It will, however, remain a staunch friend to Israel and play a role in sustaining Israel's qualitative edge.

The United States will continue to be exposed to Middle East turmoil affecting global oil markets, and therefore would seek a broader international approach to regional stability, one that would pressure other nations more dependent on Middle East oil, such as China and India, to contribute to defending free passage of oil through the Strait of Hormuz and other "global commons" concerns. The United States

[18] Tim Worstall, "Trump Doesn't Need to Bother Apple About Manufacturing in America—Foxconn Replaces 60,000 With Robots," Forbes, May 27, 2016.

will continue to maintain friendly relations with all of its allies, but it cannot continue to be a supplicant to deeply flawed, poorly governed regimes in exchange for counterterrorism cooperation, U.S. basing rights, and so on. Its relationships with Middle East "frenemies" should become more transactional. Short-term counterterrorism concerns must not always trump broader U.S. values and the longer-term agenda.

At the same time, the United States should not be under any illusion that it can "promote" democracy in the Middle East. By having more political distance from its most problematic allies, it may be more able in the future to anticipate the inevitable next round of uprisings against malgoverned Middle East regimes, whether these occur a year or decades from now. Meanwhile, the United States should step up cultural and economic exchanges with the peoples of the Middle East, hedging against the uncertain future of their governments.

Defense and Defense Spending

Assumptions

U.S. national security ultimately derives from prosperity and in the future will depend on access to and stability of the most dynamic regions of the developing world. Countries that once figured into the U.S. strategic calculus mainly as Cold War proxies have grown into economic powerhouses, with large populations, growing militaries, and regional ambitions. American prosperity will depend on peace and security in these regions, safe navigation and transport of goods and people to the important new regions, and thus on U.S. security cooperation with a range of new partners.

Collaborative security arrangements in Europe—and, potentially, in Asia—will be increasingly important; Europe, Japan, and South Korea—the nations that have most benefited from such arrangements in the past—are more than capable of contributing more to regional defense. Success in this strategy would require the United States to find new ways to assist its allies and persuade them to do more. For NATO members, this means not only achieving the agreed target of spending

2 percent of GDP, but also acknowledging that Europe is now more than capable of organizing collective defense against a belligerent but much weaker Russia. Wealthy Japan and South Korea must share more of the security burden in Asia, but must also not be made to question the credibility of the U.S. nuclear umbrella that has underwritten their security. The United States would be extremely cautious about taking on new military commitments. But it would leave no doubt about its intention to meet its obligations to existing allies—and expect more from the allies in return.

This would free up the United States to help address emerging new challenges—from piracy to public health, proliferation of weapons of mass destruction to environmental degradation—in regions that have not received much U.S. attention or resources and that will be increasingly important in the decades to come. This might include dynamic regions of Asia, Latin America, or southern, western, and eastern Africa.

Unlike the "Indispensable Nation" posture, here the United States would conceive of its contribution to global security more as a "wholesaler" than a "retailer." It would employ military force within clear limits of scope, primarily to defend allies and partners and to destroy or recapture enclaves or territories held by terrorist forces. It would avoid deploying ground forces in places like Syria, not only because of the cost but also because large U.S. deployments appear to undermine the requirement for compromise necessary to achieving long-term political solutions. Therefore, when the United States does intervene militarily, it would seek to make local leaders "own" their problems by limiting U.S. "boots on the ground."[19] It would instead contribute air, naval, and Special Operations forces where necessary. Even so, the current global security outlook is too unstable to allow spending to fall below 2.5 percent of GDP, as the Obama administration had hoped.

This strategy assumes that the most existential security threats today are supranational threats that will require collective responses: preventing the proliferation of nuclear and biological weapons, including by tightly coordinated global economic sanctions; managing the Syrian refugee crisis so that it does not destabilize Europe; establishing

---

[19] Former RAND researcher Kim Cragin is a thoughtful proponent of this view.

norms for state behavior in space and cyberspace; international efforts to counter sectarian political warfare, of which terrorism is only one tactic; coping with global pandemics; and mitigating climate change. Most of these problems are intractable precisely because they pose international *political* problems in addition to financial, technical, or military challenges. The United States should therefore focus primarily on increasing its political strength, through stronger relationships with allies, new partners in dynamic regions, and more economic clout.

While the restraint option would abandon the two-war posture and the "Indispensable Nation" option would retain it, this "Agile America" concept would pursue a minimal two-war posture. The United States would retain the ability to project substantial military power to more than one region but would rely heavily on its partners to provide basic ground and air capabilities in time of conflict. In return, it would support regional alliances and partnerships and offer contributions of air, naval, and Special Operations forces for regional operations as needed.

The size of the Army could decrease modestly, as large ground-troop commitments in two theaters would not be feasible, but the Air Force, Navy, and Marines would remain at current strength. Readiness would be "tiered," meaning not all units would be fully trained at any given point in time. There would also be greater reliance on the National Guard and Reserve. The number of Special Forces would be increased for use in counterterrorism and other operations. Access to space and a vigorous cyber defense would be essential.

Modernization of nuclear weapons would be staged accordingly. The sea-based nuclear deterrent would be updated but the land-based components of the triad would not be modernized. Development of flexible, nonnuclear strategic options would be a priority.

Climate Change

Assumptions

At the very least, climate change poses a substantial economic risk against which the United States needs to hedge, and economic incentives must be deployed to this end. The United States would aggressively promote energy R&D and U.S. commercial interests in innovative energy technologies across all major sectors of the economy. Adaptation actions described in the restraint option would apply here but mitigation actions would be added to the mix. A carbon tax or other market-based mechanisms could be implemented to promote innovation and drive low-carbon technology costs down. Tax incentives for carbon sequestration might also be considered. While the United States would be an active player in the UN Framework Convention on Climate Change, it may rely most heavily on a bottom-up process of harmonizing climate policies among countries already seeking reductions, rather than focus on a global consensus on controlling emissions.

Constraints

Such a redefinition of American national self-identity could be politically fraught. For example, the Obama administration's effort to deal with political chaos in Libya through collaborative efforts was attacked after an aide was reported to have called it "leading from behind."[20] Attempts to collaborate with China on cybersecurity or to stress diplomacy rather than the use of force in Syria have elicited accusations of U.S. weakness. Negotiations with such adversaries as Russia and Iran on issues of shared interest will remain subject to attack unless tangible benefits accrue. Differences in U.S. public opinion on climate change, particularly related to taxation or tax-like proposals on carbon consumption, could constrain action.

Risks

This strategy would aim to reduce risk by investing more effort in anticipation and resilience to insure against low-probability/high-consequence outcomes. It fails if domestic economic growth is not

[20] The term was attributed to an Obama "adviser" in Ryan Lilla, "The Consequentialist," *New Yorker*, May 2, 2011. The White House insisted it never used the phrase.

substantially improved. Absent tangible economic benefits to middle-class Americans, public support for an economic engagement policy collapses in favor of a more isolationist or more assertive U.S. foreign policy. Other risks include:

- China takes advantage of a more accommodating U.S. posture to push further. China stabilizes its economy and grows more assertive, threatening freedom of navigation in the Pacific and thus a key source of U.S. economic growth and strength.
- Allies do not step up defense spending, exposing themselves and the United States to increased military risk. Emboldened by the successful annexation of Crimea, Russia destabilizes other nations that have large Russian-speaking populations, and a weak Europe is further fractured.
- Allies' weak economies make it politically impossible to meet their other international commitments, including perhaps their pledges on climate change or humanitarian contributions.
- Absent a big military stick, the United States becomes more reliant on economic coercion, which cannot succeed without support from the rest of the world. Overuse of sanctions could undermine the financial system or create incentives for alternative systems.
- The United States lacks sufficient power projection capabilities to deal with catastrophes in distant places, whether conflicts or natural disasters.
- This strategy sets up some tension between economic growth and reducing risks of climate change, particularly if technology does not rapidly reduce the costs of emissions reductions and if global growth continues to lag. This strategy could leave the U.S. and world economies vulnerable to major disruption if the impacts of climate change are high (more than 5 percent of GWP per capita) and "bottom-up" international engagement does not reduce emissions sufficiently quickly.

CHAPTER TEN

How to Choose

Choosing the most promising strategic approach for engaging the world is neither easy nor obvious. Because many contours of the evolving global landscape are clouded, and because recent U.S. foreign and security policy experience has been mixed, Americans are now much less clear about their country's role in the world. Regaining a sense of overall direction will be essential in building long-term support for various choices and directions. Shocks, of course, can lead to abrupt changes, as the country experienced in the aftermath of 9/11 and the 2008 financial crisis, but these effects need not be permanent. Even if broad goals are established and accepted across the American polity, the paths taken are likely to veer from one direction to another, much like U.S. policy during the Cold War.[1]

Therefore, combinations of choices, tactical opportunism, and even policy reversals are every bit as likely as consistency. After all, it was the liberal Franklin D. Roosevelt, a communist foe, who engaged Josef Stalin to help defeat Adolf Hitler's Germany; Nixon, a decided anticommunist, who conceived of splitting the Soviet Union and China; and Ronald Reagan, also a decided anticommunist, who, at the 1986 Reykjavik summit, negotiated with his Soviet foes for the possible elimination of nuclear weapons. A veering and blending of paths is, in fact, most likely.

[1] For a good depiction, see John Lewis Gaddis, *Strategies of Containment: A Critical Appraisal of American National Security Policy During the Cold War*, New York: Oxford University Press, 1982.

The previous chapters explored three plausible paths that the next President and Congress might take. Our intention in developing these alternatives was not to suggest that one was better than the others but rather to illuminate the choices facing an ambivalent and divided U.S. electorate and to show that coherent foreign policy requires consensus and a willingness to back whatever strategy is chosen with the resources required to succeed. The underlying judgment regarding America's role in the world—and operating assumptions surrounding each option—vary widely.

In this chapter, we explore various ways to think about choosing among strategies. We offer six criteria that might guide decisionmaking, although no single consideration will be determinative. American values have a strong influence over policy but hardly dominate it. U.S. interests evolve over time as relationships and circumstances change. Commitments may become more or less onerous to fulfill. Technology can tilt the playing field or level it. And, of course, U.S. politics should provide course corrections over time, although the current polarization makes it much harder to form an enduring consensus on any particular issue.

What is missing is a consensus on a larger agenda that would guide the United States and its partners toward a world that is safer and more prosperous, an agenda that enables the country to take early action to prevent crises, puts forth America's strongest attributes, and can be embraced by friends and partners throughout the globe. As Dobbins noted in the initial volume in this series:[2]

> The most successful eras of American statecraft have been periods of construction: the birth of new institutions, the reconstruction of shattered nations, and the establishment of new norms for international behavior. The United States needs to combine its defense of existing institutions and norms with a rededication to such a positive agenda and to commit itself to providing the necessary resources.

---

[2] Dobbins et al., 2015, p. 119.

Here, we outline the factors we suggest the American public and leaders should consider when evaluating policy alternatives for building a positive global agenda. No foreign policy will be without critics or failings—circumstances in the world change and adapt, now more rapidly than ever—but a basic set of criteria can help inform the public's judgment on the alternatives that their political leaders have to offer.

Scrutinize Assumptions

All policy is based on a set of assumptions regarding the state of the world and the U.S. role in it. Prior to the late 19th century, U.S. leaders did not view the United States as a global power and, therefore, did not feel compelled to respond to global events. In the aftermath of World War I, over the objections of President Woodrow Wilson, the United States largely withdrew from the global scene on the assumption that its greatest strengths and future prosperity were to be found by focusing on the many opportunities at home. In the aftermath of World War II, U.S. leaders were determined not to repeat the mistakes following the previous war and instead focused on establishing a new global order with the United States firmly in the lead. This generally enjoyed widespread support that lasted roughly 60 years.

Depending on one's perspective and the assumptions that surround it, any number of paths could be chosen in support of U.S. interests. If one believes that a sound and vibrant economy is the primary source of U.S. strength, that Americans are safer today than they have been over much of their recent history, and that many of the world's toughest problems—violent Islamist extremism, for example—will not be resolved by large-scale U.S. military involvement, then a policy of domestic renewal and international restraint will be preferred. If one believes that security and prosperity are deeply intertwined, that U.S. security interests are shifting and today's most pressing problems are different from even the recent past, and that the United States benefits from globalization but must retool to adapt to its effects, then a policy that would boost U.S. agility and competitiveness will be preferred. If one believes that prosperity derives from security, that the world's problems will not be resolved by inaction and withdrawal, and that the United States and its allies and partners benefit from the world order they so painstakingly built over seven decades, then a policy of global leadership will be preferred.

What is important here is that assumptions matter, and they need to be scrutinized carefully. They establish the basis for policy and form the prism through which global events will be interpreted. Once in place, they do not change easily short of the inevitable shocks or surprises that occur. Yet a wide variety of cognitive and informational biases can distort our assumptions. As discussed in Chapter Five, humans seek information that confirms their existing views and biases. We engage in wishful thinking and confuse this with positivity. We value our present circumstances more than our future well-being. We tend to conflate what we perceive as most urgent with what is most important and what we don't know with what we consider to be unlikely.[3] Assumptions about likelihood are notoriously fallible, of course, leading to the saying in intelligence circles that "He who controls the assumptions controls the policy." Policymakers know that faulty assumptions are the cause of many dangerous blunders; in this period of intense partisanship, they should redouble their efforts to separate ideology from assumptions, the unknown from the unlikely, the truth from the tendentious.

When it comes to foreign policy decisions, the American public should scrutinize the evidence on the public record and question the assumptions that underlie courses of action being proposed by national leaders. This volume has attempted to review the evidence and unpack the cluster of assumptions that underlie the three strategic postures. It is the responsibility of the public to consider the trade-offs, risks, and opportunities associated with each, and make their decisions accordingly. We urge them to apply the "common sense test" by seeking evidence from a range of nonpartisan sources, looking for discrepancies between that evidence and the assumptions of their political leaders, and demanding explanations from their leaders about how the policies being advocated are likely to achieve the goals that matter most to them.

---

[3] Chang, 2016.

Seize Opportunities

We have argued that America's opportunities exceed its limitations. We highlight a number of opportunities, and the investments that might be needed to realize them.

Trade and Investment

As discussed in Chapter Four, the evidence that free trade benefits the United States as a whole is persuasive, as is the evidence that much more must be done to help Americans who are being hurt by the historic economic transformation now under way. With the dual goal of liberalizing trade and maintaining the legitimacy of the WTO, the United States should aim to restart a new, broad-based multilateral negotiating round. There is great value in continuing to develop rules that apply to all countries. The United States also has the opportunity to complete and approve some version of the broad trade and investment agreements for the Pacific (whether the TPP or a modification thereof) and for the Atlantic (ideally including Canada, the United States, and Mexico in a unified agreement with the EU). With its advantages in the "knowledge economy" and foreign direct investment, the United States can particularly benefit from the opportunity to help shape global investment and service trade rules.

The Western Hemisphere

With the fading of many revolutionary movements in Central and South America, senior U.S. policymakers have focused little attention on the region, apart from Venezuela. Nevertheless, long-term U.S. engagement in Colombia and Honduras have helped produce positive outcomes in each country, including long-sought peace in Colombia. With the passing of Fidel Castro, restoration of diplomatic relations with Cuba may, over time, move the region past the toxic legacy of the anti-Americanism that characterized relations in much of the last century. U.S. engagement with Mexico, Brazil, and Venezuela will be particularly determinative to the ability to make progress on migration, trade, and transnational criminal networks.

The United States could do much more to work with other countries to build a peaceful, prosperous, and secure hemisphere, extending from Cape Columba in Canada to Tierra del Fuego in Argentina. Such a goal may be more achievable than peace in the Middle East and could bring more-tangible benefits to the American people, particularly by helping Honduras, El Salvador, Guatemala, and Mexico.

Brazil's economic and political woes could provide a renewed opportunity to build closer ties with an important democracy that has a strong history of anti-Americanism.[4] Venezuela's suffering under the Hugo Chavez and Nicolas Maduro regimes could create the climate for a strategic reversal in its relationship with the United States. This could be precipitated by a change of leadership in Caracas, which should be followed by prompt humanitarian aid from Washington and ultimately a lifting of economic sanctions. Yet it also has the potential to spiral into chaos.[5]

The agreement between the government of Colombia and the Revolutionary Armed Forces of Colombia (FARC) is a positive step, though the success of its implementation is uncertain. U.S. support for a durable peace will be critical, as will international financial support for programs to demobilize and reintegrate rebels and revive communities formerly under FARC control. Economic redevelopment aid for areas whose main activity has been cocaine production could directly benefit the United States to the extent that it is paired with law enforcement efforts to halt narcotrafficking from such areas.

The Arctic

Competition is intensifying among governments and multinational corporations over control of Arctic territories[6]—and with that con-

---

[4] See, for example, Jonathan Watts and Jan Rocha, "Brazil President Weeps as She Unveils Report on Military Dictatorship's Abuses," *The Guardian*, December 10, 2014; and Sean Goforth, "Brazil's Erratic Behavior," *The National Interest*, April 11, 2012.

[5] While attention has focused on migration from Syria, we should not rule out significant migration from Venezuela to other areas in the Western Hemisphere, including the United States, should conditions there continue to worsen.

[6] Stephanie Pezard and Timothy Smith, "Friends If We Must: Russia and China in the Arctic," *War on the Rocks*, May 6, 2016.

trol, access to billions of tons of oil and natural gas, fisheries, minerals, as well as the run of maritime routes and even ownership of the seabed under the North Pole. The United States would benefit from a prompt and peaceful resolution of the competing territorial claims in the Arctic, enabling scientific cooperation and environmentally sound development of its resources.[7] Russia's defense ministry has ordered a series of new icebreakers as part of its growing military focus on the Arctic,[8] and the region has become a new locus of international espionage.[9] These developments could constrain U.S. attempts to pursue purely peaceful Arctic development unless it were to invest in capabilities to ensure access to the region. In terms of diplomacy, an activist stance on the Arctic would be constrained by congressional refusal to ratify the UN Convention on the Law of the Sea.

Africa and the Developing World

Sub-Saharan Africa hosts one of the fastest-growing regional economies, and the United States is quite popular in many African countries. The tech boom across much of the continent offers new opportunities for deeper U.S. engagement in supplying private capital. The United States is a major aid donor, but loans through U.S. aid programs, the World Bank, and IMF are typically constrained by conditionalities aimed at guarding against corruption and human rights abuses and requiring transparency. The United States might have to compete with China or other countries' sovereign wealth funds in certain types of large infrastructure lending. As the world's largest aid donor, however, the United States will continue to have the opportunity to shape how U.S. and international aid can be deployed more effectively.

---

[7] Stephanie Pezard and Abbie Tingstad, "Keep It Chill in the Arctic: Continued U.S. Cooperation in the Arctic Is Key, Despite Simmering Tensions with Moscow," *US News*, April 27, 2016.

[8] "Russia Unveils New Navy Icebreaker in Arctic Military Focus," *DefenseNews*, June 11, 2016.

[9] James Bamford, "Frozen Assets: The Newest Front in Global Espionage Is One of the Least Habitable Locales on Earth—the Arctic," *Foreign Policy*, May 11, 2015.

More broadly, the United States should also develop plans to engage the developing world in the wake of China's likely, if only temporary, retreat as its economy slows. Commodity prices have fallen precipitously, and many in the developing world that had tied their economic futures to China's unabated growth are now looking to expand their networks of economic relationships. The IMF and the World Bank expect robust growth in India, the Southeast Asian countries, and sub-Saharan Africa, although growth in Latin America and the Caribbean is projected to lag.[10] Despite such fluctuations, the developing world is expected to grow more than twice as fast as the high-income countries, creating opportunities to forge new bonds that would benefit the United States and these emerging economies.

Women

The United States has led the world in improving the lives of women and girls around the world. It is well positioned to do more, sowing good will and, in particular, helping female entrepreneurs and farmers, who are viewed by the development community as drivers of sustainable economic growth and political stability. Women entrepreneurs are estimated to have a credit gap estimated at between $260 billion and $320 billion,[11] presenting an investment opportunity for U.S. businesses and charities, but also a way for the U.S. government to engage in ways that are often nonthreatening to governments.

Global initiatives to improve women's education and public health are generally seen as more successful than other forms of development aid. The World Bank has identified girls' education as a "strategic development priority."

---

[10] Shatz, 2016, pp. 43–49.

[11] IFC Advisory Services, *Closing the Credit Gap for Formal and Informal Micro, Small, and Medium Enterprises*, Washington, D.C.: International Finance Corporation, 2013, p. 19. Sarah Kaplan and Jackie VanderBrug, "The Rise of Gender Capitalism," *Stanford Social Innovation Review*, Fall 2014. A lack of empirical research on which programs are most successful prompted USAID to commission a multiyear project to study the issue.

Energy

The revolution in extraction technologies, commonly referred to as "fracking," has already had a major geopolitical effect. The United States now ranks tenth in the world in proven oil reserves and fifth in natural gas reserves.[12] U.S. exports of natural gas to Europe could, over time, reduce dependence on Russian energy supplies. Current low world oil prices de-incentivize efficiency improvements and endanger political stability in many energy-extracting countries, such as Venezuela. Yet these developments, along with the inability of the Organization of the Petroleum Exporting Countries to enforce production quotas, offer diplomatic opportunity to revisit and potentially improve U.S. relations with producer countries. A sustained decline in oil prices would constrain the ability of such rich oil states as Russia and Saudi Arabia to underwrite a more activist foreign policy over the long term. Domestically, the money saved on foreign oil purchases could help improve the U.S. trade balance, but only if the money saved on foreign oil imports is not used to import other goods. Any breakthrough on zero-carbon energy technologies has the potential to be a geopolitical game-changer, and so deserves careful attention.

Mistakes as Opportunities

The next administration should plan for and take advantage of the multiple scenarios in which Russia, China, Iran, North Korea, or ISIS makes a serious strategic mistake. Intimidation or aggression creates an opportunity to improve U.S. security and economic relations with Russia's and China's neighbors, particularly democracies that do not wish to fall under the shadow of an authoritarian hegemon. Specifically, the United States might reassure key East Asian allies in light of China's aggressive posture in the East China Sea and the South China Sea to reaffirm America's security partnerships, offer military sales and provide assistance where needed, negotiate military access to help partners with particular problems, and position U.S. forces for possible wider regional responses. Over the last decade, there has been considerable discussion of China's investment in A2/AD capabilities—military investments designed to create

---

[12] British Petroleum, *BP Statistical Review of World Energy 2014*, London, June 2014.

a "keep out" zone in East Asia. Several of our RAND colleagues have noted that China does not have a monopoly on these capabilities and that, with U.S. assistance, important East Asian allies could create "keep out" zones of their own.[13] Even those who espouse a less active role for the United States might welcome an alternative that would enable partners to do more to provide for their own security.

The United States also could take advantage of Russia's recent provocations to forge a new NATO security agenda that can deal with the rising threat from Moscow. Close Russia watchers are divided on whether Russia has the economic wherewithal to sustain these challenges indefinitely.[14] There is little doubt, however, that NATO has an exposed eastern flank that demands immediate attention. NATO has the opportunity not only to secure its northeastern flank but also to introduce a common Euro-American agenda, which has been largely missing since the second round of NATO enlargement. For nearly two decades, NATO has been focused on security challenges outside of Europe; it is now time to recalibrate to address new challenges emanating from afar, as evidenced in the terrorist attacks in Paris and Brussels, the ongoing migration crisis, and a resurgent threat from Russia.

In addition, the United States could develop contingency plans in the event that Iran makes the gross strategic blunder of violating the 2015 Joint Comprehensive Plan of Action agreed upon with the permanent members of the UN Security Council and the EU. The existence of such back-up plans—and, more importantly, of the political will to enforce them—may be useful in maintaining Tehran's continuing compliance.

Whatever choices they make, U.S. leaders need to be attentive to such opportunities as they arise and show ingenuity, flexibility, and the ability to offer a constructive, positive alternative to authoritarianism,

---

[13] For an illustration, see Terrence Kelly et al., *The U.S. Army in Asia, 2030–2040*, Santa Monica, Calif.: RAND Corporation, RR-474-A, 2014.

[14] For an analysis of economic factors constraining Russia, including the depreciation of the ruble, state control of the economy, corruption, and the increased cost of capital due to sanctions, see Crane et al., 2016. For an overview of polling data and other political constraints on the Russian government, see "Putin's Core Support Begins to Waver," Bloomberg, June 9, 2016.

aggression, and even chaos. The United States still has a strong hand to play.

Uphold Commitments

The United States remains the partner of choice because it has made good on its commitments to other nations. This has been true since the early days of the Cold War when fledgling democracies in Europe came under assault. The United States has earned a reputation around the globe as a generous, reliable partner, from its steadfast support for its NATO partners to its long-standing commitment to Japan and South Korea; its protection of Kuwait in the face of Iraqi aggression; and its consistent record of helping nations struck by famines, tsunamis, earthquakes, and other disasters. This is not to suggest that U.S. policy is never self-interested or that its attempts to help others cannot go awry—as was the case in Vietnam and Iraq. It is to suggest that the United States is a valued partner around the globe, and that its partners look to Washington to make good on its commitments.

Some argue that the United States has carried too heavy a load and has encouraged "free riding" among many of its closest allies. The United States outspends its partners—to say nothing of its adversaries—by extraordinary sums and indeed maintains a substantial burden. There is little question that many allies can and should do more to provide for their own security. There is also significant risk that if the United States were to do less, or renege on its commitments, the result would be a more dangerous and unpredictable world. East Asia offers a relevant illustration. Some have argued that Japan and South Korea should bear more of the burden for their own security.[15] The argument has merit insofar as both nations have accumulated considerable wealth and are certainly capable of providing more for their own defense. But absent a robust

[15] This evergreen argument was made in 1989 by James Fallows and again as recently as April 2016 by U.S. senators in a meeting with the NATO Secretary General. James Fallows, "Japan: Let Them Defend Themselves," *The Atlantic*, April 1989; John Hudson, "Senators Slam NATO 'Free-Riders' in Closed-Door Meeting with Secretary General," *Foreign Policy*, April 6, 2016.

U.S. security commitment, the choices each nation might make could be disruptive and dangerous. Japan is often said to be "a screwdriver's turn away" from becoming a nuclear power and is widely deemed capable of doing so within a few months.[16] South Korea might follow, in the face of nuclear threats from North Korea or from a nuclear Japan. Japan and South Korea would have the ability to deter attacks on their own territories, but most analysts find those prospects to be anything but reassuring.

Global circumstances are now sufficiently fluid that all major international actors—friends and adversaries alike—are reevaluating their security requirements. Indeed, there is not a single U.S. ally, partner, or friend throughout the world that is not in some way recalculating its relationship with its own region and the United States. This is true in Europe, East Asia, and the Middle East. In every instance, the questions involve U.S. credibility and capabilities: Can the United States be relied upon for the long haul? Can it and will it protect its partners from attack or intimidation? Will it help deter threats as well as defeat them? Doing so requires greater reliance on denying adversaries the prospect of success, rather than punishing them after the fact, as was the premise of much of the Cold War thinking about deterrence. Denial is a difficult undertaking in today's world, but not beyond America's reach.

In Europe, being a reliable partner means making good on commitments to NATO's newest members in the face of Russian provocations. It could well mean stationing additional troops and equipment in the Baltic states and shoring up NATO's defenses along its eastern flank. As already noted, it should also include the development of a new NATO strategic concept, which has been largely missing since the days when the alliance was expanded following the collapse of the Soviet Union.

In East Asia, it means working closely with allies and partners to check China's expansion of its outward security perimeter—sometimes

---

[16] This claim is made frequently but has also been disputed by a number of experts, who say it would take Japan years and cost billions to make a useable nuclear weapon. Jeffrey Lewis, "How Long for Japan to Build a Deterrent?" *Arms Control Wonk*, blog, December 28, 2006; James R. Holmes, "Japan: Joining the Nuclear Weapons Club? It Could," *The Diplomat*, October 22, 2012.

expressed as expansion beyond the first island chain—in light of dramatic improvements in China's military capabilities.[17] The United States will need to demonstrate that it has viable counters to China's growing array of A2/AD capabilities.

U.S. credibility is now the weakest in the Middle East.[18] With the exception of Israel, America's staunchest partners were Egypt and Saudi Arabia—loyal friends, to be sure, but with troubled governing structures and disturbing human rights records. The U.S. war with Iraq changed the Middle East, displacing one of America's most challenging foes and introducing a new political and security dynamic. Iraq is now under the control of the Shia majority with little or no room for participation by the Sunni minority. This has provided a breeding ground for ISIS and has emboldened Iran, which now sees opportunities to expand its influence throughout the region. The Arab Spring swept many long-term rulers from power, including in Egypt, without the prospect of viable successors to follow. It left America's traditional partners harboring severe doubts regarding the long-term viability of their relationships with the United States. Continuing chaos in the Middle East is likely, and U.S. policy may need to be remade from the ground up. It will be difficult to imagine a stable future absent a significant and constructive role for the region's moderate Sunni populations.

In all three regions, U.S. leaders need to make good on commitments and be clear-headed about making new ones. During periods of relative stability, taking on new security arrangements is tempting and appears relatively cost-free. NATO enlargement offers a good illustration. The enlargement of the alliance enjoyed popular support as a way of extending and anchoring the democratic peace that swept through Europe at the end of the Cold War. It also appeared to be relatively cost-

[17] For an excellent depiction of China's growing military prowess along with its limitations, see Heginbotham et al., 2015.

[18] Lack of U.S. credibility in the Middle East is a rare point of bipartisan agreement, although critics of the administration differ on its causes. See, for example, Nicholas Burns, "Barack Obama Has Risked U.S. Credibility in the Middle East," *Financial Times*, March 22, 2016; and Musa al-Gharbi,"Why America Lacks Credibility in the Middle East," *Foreign Policy in Focus*, March 10, 2015.

free. Russia was judged as no longer posing a serious threat to NATO, and Europe was judged to be stable and secure. A frequent expression over the years among close NATO observers was "out of area or out of business," indicating a belief that NATO needed a common security agenda that allowed it to act beyond Europe's boundaries to deal with larger global challenges. NATO's long-term commitment in Afghanistan was seen as a promising sign.[19]

Rising trouble could be seen in Russia's 2007 cyberattacks on Estonia and its 2008 invasion of Georgia. But not until Russia's 2014 annexation of Crimea and its war against Ukraine did the true cost of NATO enlargement become clear.

The question here is not about the wisdom of expanding the alliance. The lesson is about being clear-headed regarding the implications of new commitments as they are being considered. The first round of NATO enlargement, which brought Poland, the Czech Republic, and Hungary into the alliance, occurred after a vigorous national debate. The second round of enlargement, which brought Bulgaria, Estonia, Latvia, Lithuania, Romania, Slovenia, and Slovakia into the alliance, occurred with little to no debate. Any such future commitments, in Europe or elsewhere, need to be thoroughly evaluated because once the United States makes commitments to partners, it needs to be prepared to back them up—with force, if necessary.

Play Both the Long and Short Games

U.S. leaders have been most successful when they have been able to distinguish the challenge of the moment from the fundamental policies and structures needed to produce a more secure and prosperous future. Even while fighting World War II, Roosevelt and his advisers were contemplating and building the institutions and policies to cope with the world to come. Although their visions had to be adapted as

---

[19] For a discussion of the political circumstances surrounding NATO's involvement in Afghanistan, see Andrew R. Hoehn and Sarah Harting, *Risking NATO: Testing the Limits of the Alliance in Afghanistan*, Santa Monica, Calif.: RAND Corporation, MG-974-AF, 2010.

the Cold War took shape, they conceived of many of the basic institutions and frameworks that survive to this day. Nixon's opening to China was aimed at changing both the long and short games, seeking long-term leverage over the Soviet Union even while trying to end the war in Vietnam. Reagan played the short game in negotiating arms control agreements with Soviet leader Gorbachev while also trying to bring about a Soviet collapse. Both strategies worked. Clinton sought to enlarge the democratic peace by inviting those once captured in the East-West divide into the institutions of the West. Clinton's moves were broadly popular at the time but almost certainly sowed the seeds of some of the challenges the United States confronts today.

George W. Bush sought to transform the Middle East by displacing Saddam Hussein in Iraq and ushering in new democratic forms of governance. What followed was more than a decade of war and a hardening of Iraqi and Middle East politics. The long-term consequences will not be known for years.

Obama sought to end the wars in the Middle East and pivot to Asia, another long-game strategem, but the move lost momentum because the short game in the Middle East was far from over and new strategic challenges arose in Europe. Indeed, ending the U.S. troop commitment in Iraq while the political and security situation there remained so fragile contributed, in part, to the rise of ISIS—a short-game decision with long-game consequences.

Finally, every President since Clinton has had to contend with the basic breakdown in global governance. From Mogadishu to Mosul, Aleppo to Benghazi, the breakdown in governance has proven to be a profound challenge to global security. Al Qaeda was able to plot its attacks against the United States amid the chaos that was the Taliban's Afghanistan. While most Americans eschew the notion of being the world's policeman, there needs to be a wider public recognition that the world's ungoverned and misgoverned places can provide refuge to the most dangerous terrorist actors. The U.S. government well understands that governance is an essential element of stability; for example, in preventing the resurgence of al-Shabaab in Somalia.[20] Reestablishing effective governance in many key parts of the world is a long-game interest of the United States—one that will likely remain politically unpopular in an era of fiscal strain, but must nonetheless be pursued.

Align Interests and Values

Much has been written about the importance of American values in the making of U.S. foreign policy[21]—how interests and values ultimately need to align, and yet the occasional necessity for interests to supersede values in times of peril. It was Roosevelt who said, "My children, it is permitted you in time of grave danger to walk with the devil until you have crossed the bridge,"[22] as a signal to the partnership he would need to forge with Stalin's Russia in order to prevail in World War II. U.S. presidents have made similar arguments since that time—to justify relationships with dictatorial and corrupt regimes during the Cold War, to maintain access to strategic locations, to ensure availability of needed energy and commodities, and more recently to explain some elements of the counterterror coalition that emerged after 9/11 and continues to this day. Yet the appeal to values for political support is well established in American political life; it is the basis of America's claim to "exceptionalism" in world affairs and an affirmation of our national identity.[23]

Interests and values do not always align, certainly not in the short term. When they do not, policy will be forged to find ways to improvise, even as policymakers chastise or work behind the scenes and face criticism from many quarters. This is particularly true of U.S. democ-

[20] Seth G. Jones, Andrew Liepman, and Nathan Chandler, *Counterterrorism and Counterinsurgency in Somalia: Assessing the Campaign Against Al Shabaab*, Santa Monica, Calif.: RAND Corporation, RR-1539-OSD, 2016.

[21] Henry Kissinger, *Years of Renewal*, New York: Simon and Schuster, 1999, p. 1073; Kagan, 2012; Chester Crocker, "The Strategic Dilemma of a World Adrift," *Survival*, Vol. 57, No. 1, February–March 2015.

[22] Gaddis, 1982, p. 3.

[23] Kissinger, 1999, p. 1073; Kagan, 2012; Crocker, 2015.

racy promotion, and of its international engagement on human rights, articulated by Roosevelt in his "Four Freedoms" speech, institutionalized in the U.S. government under Carter, and pursued with more or less vigor by Presidents since then.

Debates over the tension between interests and values were once prevalent regarding U.S. policy in Latin America, parts of Europe (recall Spain under Francisco Franco), East Asia, and certainly the Middle East. The very rapid rise of democracy across the globe that came in the aftermath of the Cold War caused a decline in the tensions in U.S. bilateral relations with many countries. Still, recent trends will likely continue to pose severe challenges to U.S. policymakers in the decade to come.

The authoritarian regimes that have survived are the most experienced, effective, and well-defended. These autocrats and monarchs have endured wave after wave of global democratization, withdrawal of support from their Cold War sponsors, decades of U.S. democracy promotion efforts, and global criticism of their human rights records. Some will probably indeed fall over the coming decades, but others may endure. New harsh Islamist states may arise.

Given the disastrous outcomes of all but one of the "Arab Spring" democracy movements,[24] the United States will be reluctant to push its democratization or freedom agenda in countries where regime change seems likely to bring even more brutality or chaos—which may include the majority of nondemocracies. Future administrations might instead focus on how to support the efforts of the emerging and developing democracies to improve their governance as well as their economies. Some argue that the fate of the liberal democratic order will primarily be determined not by the struggles of the Middle East, but by the success or failure of the democratic model in such rising powers as Brazil, India, Indonesia, Turkey, and South Africa.[25] Some of the smaller or more fragile ones—Burma, Tunisia, Nigeria—may require U.S. assistance to survive.

---

[24] The exception is Tunisia, where an Islamist government handed over power to an elected rival in 2014.

[25] Piccone, 2016. Tunisia, Burma, and Nigeria may also be considered critical bellwether states.

On human rights, progress is typically measured in decades, not presidential terms. It is achieved variously through changes in government, international public embarrassment, private diplomatic engagement, and most importantly, domestic political pressure. Both progress and regression in democracy tends to occur quite suddenly, albeit after years of gestation, and these tipping points have defied efforts at prediction.

U.S. engagement on human rights is typically polarizing but since 1976 has led to unprecedented advances in rights and governance around the world. U.S. effectiveness can be undermined by partisanship, as rights-abusing regimes try to "wait it out" in hopes that the next administration will more sympathetic or seized with other problems. Therefore, the next administration, whatever its political views, would benefit from continued active participation in international forums—perhaps even the problematic UN Human Rights Council, which is now headed by Saudi Arabia. The incoming administration will be greeted with a list of human rights and humanitarian problems that the United States "should" address, yet it must set priorities. Which issues deserve high-level attention, and why? And where will the United States choose to "walk with the devil"?

Naturally, each President will make his or her moral judgments, which may or may not reflect the views enunciated as a candidate. From the standpoint of strategy, however, three criteria should be applied to prioritizing the promotion of American values.

First, does the problem strike at the heart of American identity? By this criterion, the United States has a strong national interest in leading global efforts to address refugee and migration policies; disrupting human trafficking in global supply chains (now referred to by the Secretary of State as "modern slavery"[26]), and improving compliance with the international conventions banning torture, child soldiers, and rape and slavery as weapons of war. However difficult and

[26] U.S. Department of State, *Trafficking in Persons Report*, Washington, D.C., July 2015. The State Department report focuses on the problem of "modern slavery" in international supply chains, as well as the use of slavery by ISIS and others as a tactic in armed conflict.

fitful progress on these issues may be, they reflect universal values that are broadly shared among U.S. citizens and allies.

Second, the policy community should focus on initiatives that are likely to make a difference. That is not to say the United States should abandon its principles when confronted with obstacles, but it is to say that some problems are riper for resolution than others. It is also a way of recognizing that most issues are wrapped in layers of complexity that do not lead to simple or easy resolution. It matters whether a decision to "walk with the devil" in fact produces tangible or immediate benefits. For example, even proponents of the nuclear deal with Iran criticized the Obama administration for not linking the final deal to the release of U.S. prisoners held in Iran, although the Americans were subsequently released on the day the United States lifted economic sanctions. The American public, especially the informed public, understand negotiating with competitors and adversaries, but they expect those negotiations to be undertaken from positions of strength and to achieve tangible benefits.

A third consideration is whether human rights policy can be crafted in such a way as to support, rather than undermine, broader U.S. foreign policy goals. (Sanctions against Putin's Russia are a good example.) Judged by these three criteria, human rights issues of major strategic importance include:

1. **Middle East refugee and migration crises.** This and related immigration issues have already become a subject of intense debate among the American electorate, yet the debate has yielded little by way of insight or policy direction. The United States, which has for more than a century billed itself as the beacon of hope for the downtrodden and as the supplier of global leadership in establishing refugee policy, faces reputational risk if it leaves Europe and others to lead efforts to save and support refugees. Moreover, the destabilizing effects of the largest refugee crisis since World War II, including the long-run effects of a generation of children not attending school, are linked to violence and therefore likely to have strategic import. It is already clear that the Syrian refugee and migration crises are threatening parts of Europe, could well displace

long-standing European leaders, and may lead to the collapse of the EU's open border policy—the Schengen area—which would not only be a setback for EU policy but could contribute further to a downturn in the global economy. Addressing the crisis need not necessarily lead to thousands seeking refuge in the United States, although it may imply a more active U.S. role in the Syrian and Libyan civil wars to help alleviate the situations that have forced so many to flee in the first place.

2. **corruption.** Corruption and other failures of the rule of law are often the underlying cause of countries' inability to make progress on their own human rights and humanitarian problems, as well as to develop thriving economies. Over time, reducing or eliminating corruption not only brings coherence and order to the societies being served but also helps create economic opportunity for the U.S. financial and commercial sectors that are not permitted to use bribery as a tool of commerce (as some competitors do) and who benefit most from a rules-based environment in doing business abroad.
3. **support for civil society.** As already mentioned, the world's surviving autocrats have become more astute in devising ways to cling to power, not only by crushing opponents but also through a range of activities aimed at intimidating civil society, from surveillance and censorship to legal harassment and extrajudicial punishment. Delegitimizing nongovernmental organizations by branding them as agents of a foreign power is a common technique.[27] Nevertheless, the United States (as well as U.S. nongovernmental organizations, businesses, and private citizens) can continue to support civil society groups through diplomatic and assistance efforts,[28] people-to-people exchanges, and engagement among experts. Particularly when political reform seems unlikely if not impossible, the United States can play a critical role through the patient cultivation of ties with civil society groups.

---

[27] Freedom House, 2016a.

[28] USAID, Center of Excellence on Democracy, Human Rights, and Governance, *Stand with Civil Society: Best Practices*, January 2014.

Limit Regret

At a minimum, U.S. leaders have a moral requirement to protect their people from serious harm. It has been said the 9/11 attacks were the first large-scale external attacks on American soil since the British invaded and burned Washington during the War of 1812.[29] These have been followed by a growing number of high-casualty internal attacks by so-called "lone wolves," including U.S. citizens inspired or justified by violent Islamist ideology, but so far without the use of nuclear, radiological, or biological weapons.

There is little doubt that the responsibility to protect the American people weighs heavily on every U.S. president. In the early days of the republic, this meant defending U.S. borders and coastal waterways against invasion and incursion, protecting against natural disaster, and, in time, establishing a public health system to protect against pandemic and plague. These basic requirements remained in place until the dawn of the nuclear age.

During the Cold War, protecting the American people from serious harm meant all of the above plus a daunting new requirement: precluding nuclear attack on the United States. This led to the development of theories of deterrence (discussed in Chapter Five) and enormous investments in the intellectual makeup and arsenal of deterrence. This was a monumental effort and proved to be a much riskier effort than any at the time understood or many choose to remember.

For more than a decade following the end of the Cold War, the United States experienced a relative sense of invulnerability. U.S. interests were challenged abroad, to be sure, but Americans felt relatively safe at home. U.S. borders were secure from invasion and attack, and the risk of global pandemic had not yet crept into America's consciousness. Nuclear weapons continued to exist, although attention was moving away from the threat of an all-out attack and more toward concerns about the possibility of "loose nukes," either in the hands of a rogue state or possibly a terrorist. Although some worried about the dire consequences of a nuclear or biological attack, this was not at the forefront

---

[29] Hawaii was a U.S. territory but not yet a state in 1941 when Japan attacked Pearl Harbor.

of American concerns. If the world continued to experience problems and turmoil, most were perceived to be kept safely at a distance.

The 9/11 attacks brought home the threat of terrorism that remains uppermost in the American consciousness to this day.

In response to terrorist threats, both the Bush and the Obama administrations resorted to military responses and neither found effective ways to engage the ideological struggle.[30] The Bush administration committed substantial forces in Afghanistan and Iraq but left the country reeling with the sense of never-ending wars—and the sentiment that the United States is not particularly good at fighting these types of wars. The Obama administration bought time with the 2015 nuclear agreement with Iran. The issue, however, is at best deferred and hardly resolved. Future presidents will almost certainly be contending with Iran's nuclear and political ambitions and all that they portend for the highly volatile region. Concerns about a nuclear proliferation cascade throughout the Middle East—first Iran, followed by Saudi Arabia, Turkey, and even Egypt—have been put on hold for now but not for all time.

The same holds true for North Korea and its nuclear ambitions. For decades, the long game has been to keep North Korea in check in expectation of its ultimate collapse. The regime that many said could never survive has proven to have exceptional longevity. Left unchecked, North Korea's nuclear and missile programs not only present an immediate threat to its neighbors—and, ultimately, the United States—but could also have cascading effects throughout East Asia, as South Korea and Japan contemplate how to respond to this new security dynamic, even seeking nuclear arsenals of their own should they lose confidence in the U.S. nuclear umbrella. While the United States and China share a strategic interest in limiting North Korea's ambitions, progress has been thwarted by Beijing's more immediate concern about preventing a collapse of the North Korean government and a resulting migration crisis on its borders.

---

[30] Angel Rabasa, "Where Are We in the 'War of Ideas'?" in Brian Michael Jenkins and John Paul Godges, eds., *The Long Shadow of 9/11: America's Response to Terrorism*, Santa Monica, Calif.: RAND Corporation, MG-1107-RC, 2011.

The hopes that Russia would join the larger zone of democratic peace have been dashed for now. President Putin is playing a particularly aggressive, dangerous game in trying to reinstall a buffer along Russia's periphery. Given the state of Russia's economy, to say nothing of its demography, it is possible that Russia may have found the limits of its reach. But there is abundant danger in Russia's ongoing probing, including, perhaps, by interfering in the U.S. political process, and it remains one of the few powers that is capable of threatening the United States directly. As we already noted, making good on U.S. commitments to the newest NATO partners will be an important obligation for the next administration. This should be done in a way that does not encourage Russian adventurism elsewhere.

Limiting regret with respect to climate change represents one of the biggest gambles of all. The consequences on national and international economic and social well-being for generations to come could be staggering. The next President and Congress will decide whether U.S. leadership means taking a cautious wait-and-see approach to the science of climate change as national policy; muddling along with adaptation actions and hoping other countries bear the burden of emissions reductions (just in case); or striking an aggressive position on adaptation and massive investment in decarbonizing the economy to reduce greenhouse gas emissions as well as sequestration should it prove necessary. As already discussed, the vast range of uncertainty over the costs of action and inaction makes decisionmaking especially difficult. The next President and Congress will soon be placing their bets, but they would be well advised to consider the benefits of an adaptive but still bold approach to tackling a problem whose odds will only grow longer.

Whatever the path chosen, U.S. leaders will want to demonstrate that they have taken all reasonable efforts to limit regret when it comes to securing the American people and territory from serious harm. It is their duty to protect the country, even against unlikely events, when the consequences could be grave and irrevocable. The American people deserve a government that has clarity on potential threats and a credible plan of action.

CODA

The Challenge of Leadership: Aligning Vision, Values, Interests, and Resources

This volume—indeed, the entire Strategic Rethink project—has been organized to answer the question, "What is America's role in the world of the early 21st century?" and to see if there is a guiding concept—a "grand strategy"—that would give direction and coherence to our diplomacy, defenses, and economic engagement in a turbulent world.

The other five volumes of the project, along with this overview volume, have explored in considerable detail the many challenges and policy choices of this new era in world affairs. We have outlined three alternative strategic concepts, each of which is reflected in the contemporary domestic political debate. Each has its own rationale for meeting today's diverse challenges; each has its own objectives and priorities, resource and financial requirements, and degree of risk to national interests.

It is clear that the complexity of contemporary international relations does not lend itself to any simple strategic concept or statement. In the early years of the Cold War, in George F. Kennan's time,[1] the existential security threat from the Soviet Union focused attention and resources around the strategic concepts of containment and deterrence. Amid vigorous policy debate and impelled by the security crises in Berlin and Cuba, there developed a measure of coherence and alignment between values (anticommunism and the institutions of the free world), and countering the Soviet threat to U.S. and allied security (a nuclear and conventional deterrent and a global network of alliances)

---

[1] George F. Kennan was the State Department official who first proposed the policy of containment in an anonymous Foreign Affairs article in 1947.

that gained public support and sustained the mobilization of national resources needed to fund robust defenses. There was substantial alignment among vision, values, national interests, and the resources needed to promote an effective foreign policy.

Today, our concepts of threat and opportunity are misaligned with policy and purpose, as is evident in our growing domestic polarization. We value freedom and opportunity, yet substantial segments of the public no longer perceive "the American dream" as attainable. We demand security, yet have not reconciled this fundamental requirement with the equally valued objective of privacy. We are confused as to how to organize and fund our diplomacy and defenses against threats ranging from nuclear proliferation, terrorism, ethnic turmoil, political instability, cyberattack, climate change, and pandemics. We seek economic growth but are uncertain whether international trade benefits or degrades standards of living. Resolving these policy dilemmas and building a national political consensus is the primary responsibility of—and challenge to—our leadership. Yet political polarization impedes reconciling domestic social and economic needs and the long-term fiscal threat of the national debt with external challenges to our security and economic well-being.

This fundamental policy challenge requires future presidential administrations to develop a strategic concept that can manage these dilemmas, clarify threats and opportunities, assign priorities, and give a certainty of purpose to our role in the world. Whether that concept stresses domestic renewal, a refocusing of priorities, or activist international leadership will ultimately be decided by the American electorate. And, in practice, it is likely to be some combination of the strategic alternatives detailed in this volume.

The United States has great national strengths to bring to that effort. The next administration's challenge—any administration's challenge—is to align a vision of our role in the world with fundamental values and national interests, so as to gain the political support needed to break the domestic political deadlock that prevents adequate resourcing of both domestic and foreign policy needs.

Without such a concept, our international engagement will be ad hoc and reactive; it will lack the coherence to be effective and gain the necessary political support. Without political support, our diplomacy and defenses will be underresourced, heightening risks to our security. And in a world in turmoil, there is the risk that we will suffer another 9/11 or worse—a "strategic surprise" or system shock to our security or economy—and respond in ways that divert us from focusing resources on our fundamental national interests.

In the end, America's greatest strength is its ability to develop and maintain friends and partners across the globe. No other competitor has this advantage; to surrender or squander this advantage would be a catastrophic strategic failure. Maintaining the friendship and trust of America's most important allies must be the centerpiece of any strategic approach going forward.

APPENDIX

Excerpts from the Conversation on Grand Strategy

Since the end of the Cold War, there have been persisting efforts in the academic world and think-tank community to develop a strategic concept appropriate to a time of dramatic changes in world affairs. No consensus has emerged from these efforts, but certain assumptions about America's role in the world are widely shared: We are living in an era of turbulence driven by transformative political, economic, and technological changes; the United States must continue to exercise international leadership; the country's resources for supporting international activism are limited; the maintenance and strengthening of alliances and partnerships is essential to pursuing U.S. security and economic interests; and policies should be designed to promote fundamental national values.

Despite these general areas of agreement, there are significant areas of disagreement: whether national priorities and limited resources should emphasize domestic needs or international challenges; the appropriate range of U.S. activism and leadership abroad; and which challenges to our security, economic interests, and national values should be given highest priority.

The following excerpts from the conversation among national security and foreign policy thinkers about America's role in the world give an overview of much of the post–Cold War literature analyzing or commenting on strategic policy alternatives. They are not meant to be comprehensive of a voluminous body of material, but to provide a sense of the range of thinking in the foreign policy and national security community. The excerpts are presented in chronological order of publication to reflect, in some measure, reactions to the changing international environment over the past two decades.

> Noting the lack of a domestic political consensus in support of a strategic vision for America's role in the post-Cold War world, this article assesses four grand strategies competing in public discourse: neo-isolationism; selective engagement; cooperative security; and primacy. It concludes that domestic politics will likely produce an *ad hoc* approach to the world "until a crisis impels a choice."
>
> Barry R. Posen and Andrew L. Ross,
> "Competing Visions for U.S. Grand Strategy,"
> *International Security*, Vol. 21, No. 3, Winter 1996/97

> The deepest reason for America's difficulty in the 1990s with developing a coherent strategy for a world in which its role is so central was that three different generations with very different approaches to foreign policy were disputing America's role. . . . So long as the post-Cold War generation of national leaders is embarrassed to elaborate an unapologetic concept of enlightened national interest it will achieve progressive analysis, not moral elevation.
>
> Henry Kissinger,
> *Does America Need a Foreign Policy?*
> New York: Simon and Schuster, 2001, pp. 28, 31

> . . . No nation and no institution is capable of dealing with current and future problems on its own. The only way to deal with these threats and challenges is through an integrated and allied strategic approach, which includes both non-military and military capabilities.
>
> Klaus Naumann et al.,
> *Towards a Grand Strategy for an Uncertain World: Renewing Transatlantic Partnership*,
> Lunteren, Netherlands, Noaber Foundation, 2007

The international world of states . . . is where the greatest issues of the human condition are played out. A sacral nature must infuse world order if it is to be legitimate.

Charles Hill,
Grand Strategies: Literature, Statecraft, and World Order,
New Haven, Conn.: Yale University Press, 2010, p. 293

. . . The strategic narrative of the United States in the 21st century is that we want to become the strongest competitor and most influential player in a deeply inter-connected global system, which requires that we invest less in defense and more in sustainable prosperity and the tools of effective global engagement. . . . To achieve this, we will need the tools of development, diplomacy and defense—employed with agility through an integrated whole of nation approach.

"Mr. Y." (Captain Wayne Porter, U.S. Navy,
and Colonel Mark "Puck" Mykleby, U.S. Marine Corps),
A National Strategic Narrative,
Washington, D.C.: Woodrow Wilson Center, 2011, pp. 3, 13

The most basic idea of strategy is the application of strength against weakness. . . . Good strategy is coherent action backed up by an argument, an effective mixture of thought and action with a basic underlying structure. . . . It does not require one to sort through legalistic gibberish about the differences between visions, goals, strategies, objectives and tactics. . . . It is very straightforward.

Rumelt, 2011, pp. 9, 77

> Some pessimists counsel that retrenchment is a dangerous policy, because it shows weakness and invites attack. . . . Faced with diminishing resources, great powers [can] moderate their foreign policy ambitions and offer concessions in areas of lesser strategic value. Contrary to the pessimistic conclusion of critics, retrenchment neither requires aggression nor invites predation.

Paul K. MacDonald and Joseph M. Parent,
"Graceful Decline? The Surprising Success of Great Power Retrenchment," *International Security*,
Vol. 35, No. 4, Spring 2011, pp. 7, 10

> During the Cold War, grand strategy and military strategy were united by a clear enemy and explicit geographical focus, and so provided a form of continuity that ran for more than 40 years. . . . If strategy is a matter of combining means, ways and ends, what are the ends toward which a state, nation or group is aiming when it cannot be precise about the future context within which its means and ways are being applied? Answering that question is the central conundrum of grand strategy [today] . . .

Hew Strachan,
"Strategy and Contingency,"
International Affairs, Vol. 87, No. 16, November 2011,
pp. 1281, 1293

> . . . To respond effectively in both the western and eastern parts of Eurasia, America must adopt a dual role. It must be the *promoter* and *guarantor* of greater and broader unity in the West, and it must be the *balancer* and *conciliator* between the major powers in the East. . . . But to have the credibility and the capacity to pursue both successfully, America needs to show the world that it has the will to renovate itself at home.

Zbigniew Brzezinski,
Strategic Vision: America and the Crisis of Global Power,
New York: Basic Books, 2012, p. 185

. . . An American grand strategy of democratic internationalism should pursue five goals: increasing equality of opportunity, assuming responsibility, smartly managing interdependence, leading coalitions and recasting global bargains, and building the democratic community. . . . [I]t is vital for the United States and the other democracies to continue to engage the powerful nondemocracies and encourage their greater participation in the broader global liberal order.

Daniel Deudney and G. John Ikenberry,
Democratic Internationalism: An American Strategy for a Post-Exceptionalist Era,
New York: Council on Foreign Relations, working paper, November 2012, pp. 8, 19

Strategic Agility . . . seeks to avoid US involvement in protracted ground wars and emphasizes the importance of technologically superior assets that can quickly and decisively eliminate threats to the United States and its allies. . . . This study recommends management reforms, changes to military force structure, and reduced modernization costs.

Stimson Center,
Strategic Agility: Strong National Defense for Today's Global and Fiscal Realities, Washington, D.C., 2013, p. 3

By working closely with its long-standing allies—both European and Asian—the United States has demonstrated that it can act assertively across a range of issues. . . . These allies share a like-minded world view and possess a preponderance of power that has given them tremendous leverage in advancing the norms of a liberal world order. But they are missing a critical element for success: a collective institutional vehicle for strategic coordination.

Ash Jain,
Like-Minded and Capable Democracies: A New Framework for Advancing a Liberal World Order,
New York: Council on Foreign Relations, January 2013

. . . The country must put its own house in order and, with willing partners, author a prosperous, secure and sustainable future. . . . The United States must lead the global transition to sustainability [by dealing with the strategic issues of economic growth and inclusion, ecological depletion, and lack of institutional resilience]. America will have to build and strengthen capable partners to provide basic security assurances.

Patrick C. Doherty,
A New U.S. Grand Strategy,
Sacramento, Calif.: New America Foundation, January 9, 2013

This article summarizes three strategic postures put forward by the Atlantic Council—America leading "actively, vigorously, and strategically"; by the Council on Foreign Relations—"democratic internationalism"; and The New America Foundation—leading the "global transition to sustainability."

Anne-Marie Slaughter,
"Does Obama Have a Grand Strategy for his Second Term?
If Not, He Could Try One of These," *Washington Post*,
January 18, 2013

The United States must move from efforts to sustain primacy to a strategy of discrete and targeted influence that prioritizes capabilities of broad applicability and comparative advantage and relies more on frugal, indirect, and asymmetric strategies. . . . [T]he essential U.S. global strategy over the next decade must shift . . . to a role as strategic catalyst and coordinator.

Michael J. Mazarr and the NDU Strategy Study Group,
Discriminate Power: A Strategy for a Sustainable National Security Posture,
Philadelphia: Foreign Policy Research Institute,
May 2013, pp. 1, 9

. . . The United States is going to have to adopt a coherent Eurasian strategy that integrates European, Middle Eastern, South Asian and East Asia policy into a comprehensive design . . . we shall have to prioritize the repair and defense of alliances . . .

Walter Russell Mead,
"Grand Strategy: The End of History Ends," *American Interest*,
December 2, 2013

A well designed grand strategy can help a country leverage its own strengths while exploiting an opponent's weaknesses, and it can provide the sustained focus that is necessary to succeed in medium and long-term rivalries.

Brands, 2014, p. 9

The breadth and range of its allies are one of America's most important sources of power, economically and strategically. Reinforcing its allies, and standing by them in need, must continue to be central to American policy.

Jones, 2014, p. 201

A reconstruction of the international system is the ultimate challenge to statesmanship of our time. The penalty for failing . . . [will likely be] an evolution into spheres of influence identified with particular domestic structures and forms of governance . . .

Kissinger, 2014, p. 371

> . . . Presidents and their advisers have usually drawn policy answers from one of two strategic schools . . . the school of "more" and the school of "less." . . . Presidents of the "more" school generally practiced some version of . . . maximalism. . . . Presidents of the "less," by contrast, have had to oversee retrenchment. . . . As long as the world works no better than it does, America will have many reasons to try to make up the difference between mediocre results and good ones—and to rescue what it can of our maximalist past.
>
> Stephen Sestanovich,
> *Maximalist: America in the World from Truman to Obama*,
> New York: Knopf, 2014, pp. 326, 336

America's national security elites act on the assumption that every nook and cranny of the globe is of great strategic significance and that there are threats to U.S. interests everywhere. . . . They live in a constant state of fear. [But] the United States is a remarkably secure country. . . . The United States should not be the world's policeman. . . . [It should] abandon its interventionist strategy of global domination. . . . [But] there are three regions of the world—Europe, Northeast Asia, and the Persian Gulf—that are indeed of vital strategic importance to the United States. . . . It should make sure it remains the most powerful country on the planet, which means making sure a rising China does not dominate Asia. . . .

John J. Mearsheimer,
"America Unhinged," *The National Interest*,
January 2, 2014

This article explores policy alternatives beyond retrenchment or assertiveness on the assumption that policy making in a time of dramatic change is a continuing process of "recalibration to the geopolitical, economic, technological and other dynamics driving this 21st century world."

Bruce W. Jentleson,
"Strategic Recalibration: Framework for a 21st Century National Security Strategy," *Washington Quarterly*,
Spring 2014

. . . The unanticipated fall of the Soviet empire . . . raised anew the question of how to define America's purpose and its interests in the absence of an obvious threat. . . . a question of identity and purpose. . . . [S]igns of the global order breaking down are all around us. . . . Today it is America's world order that needs propping up. . . . [T]he world lacks any genuine overarching legal or institutional authority, much less a democratic authority, to which all nations can subordinate themselves. [T]here is no [other] democratic superpower waiting in the wings to save the world if this democratic superpower falters.

Robert Kagan,
"Superpowers Don't Get to Retire," *New Republic*,
May 26, 2014

Adoption of the human security paradigm as grand strategy can inclusively and powerfully integrate United States Government (USG) functions for international development, stability operations and building partner capacity. . . . [A] grand strategy founded on the dual 'freedom from fear' and 'freedom from want' aspects of human security can reflexively operationalize internal USG collaboration resulting in increased efficiency in engagement with states, non-state actors and populations at large.

Colonel John C. Anderson,
Changing the Game: Human Security as Grand Strategy,
Carlisle, Pa.: United States Army War College Press,
June 2014, pp. 3, 36

Coherent and effective political direction is the essential precondition to strategic success. . . . Enduring approaches to national security include: meeting threats as far from the homeland as possible; using a mix of tools including intelligence, diplomacy, forward presence, and economic power—reserving military force as a last resort. . . . A persistent uncertain and unstable international security environment places a premium on U.S. leadership. . . . Strategic "shocks"—unanticipated crises requiring strategic responses—will be more the norm than not.

R. D. Hooker, Jr.,
The Grand Strategy of the United States,
Washington, D.C.: National Defense University,
October 2014, pp. 21, 22, 26

The international system is adrift because there is an unregulated diffusion of authority, agency and responsibility . . . presenting the strategist with the challenge of incoherence. . . . There is an authoritarian statist backlash trending toward a "normative" international re-polarization. Collective efforts at conflict management are spontaneous and ad hoc. The problems of international disorder are systemic and can only be fixed by the system's leading powers.

Crocker, 2015, pp. 9, 13, 15, 20

. . . The duration of the American century depends upon a broad set of alliances and will increasingly do so in the new context of world politics. . . . [The United States] must shape the international environment and create incentives for others through trade, finance, culture, and institutions, and forming networks and institutions for action.

Joseph S. Nye, Jr.,
Is the American Century Over?
Cambridge, Mass.: Polity Press, 2015, pp. 123–124

U.S. strategy should focus on five primary problems—the absolute need to deter a nuclear attack on the homeland and a core of four optional scenarios—rather than attempting to support a hollow doctrine of American exceptionalism or sustain self-proclaimed leadership that has brought benefits neither to the world nor to ordinary Americans.

Shlapak, 2015, p. 75

America cannot manage global affairs by itself, owing to a smaller margin of power compared with the past. American leadership therefore must be oriented toward building effective partnerships with allies and friends. . . . [I]n order to ensure continued American preeminence, [our leaders] must also leverage the key dynamic trends that are unfolding across the globe . . . including energy, urbanization, technology, climate, individual empowerment, and communications. . . .

Barry Pavel and Peter Engelke,
Dynamic Stability: US Strategy for a World in Transition,
Washington, D.C.: Atlantic Council, April 2015, pp. v, 1

What sort of superpower do you believe the United States of America should be? . . . Indispensable America resonates because we do live in an increasingly dangerous world. . . . A Moneyball foreign policy would set priorities and stick to them, allowing Washington to devote its limited means to achieving our most important objectives with plans that are politically and financially sustainable. . . . I choose Independent America. . . . [W]hat our elected leaders have been doing for years [is] improvising responses to challenges as they arise without a comprehensive strategy to guide their choices.

Bremmer, May 2015, pp. 191–204

> . . . In the new world disorder, America needs national security policies that begin and end by asking what's in these policies for Americans, not what foreign nations long dependent on our protection might think about them. There is no reason for us to continue to shoulder burdens others can now bear. We should build our strength while holding it in reserve. We should act only when it's in our interest to act.
>
> Freeman, 2016

Abbreviations

9/11	September 11, 2001, terrorist attacks
A2/AD	anti-access/area denial
AIIB	Asian Infrastructure Investment Bank
ASEAN	Association of Southeast Asian Nations
C	Celsius
CBO	Congressional Budget Office
CIA	Central Intelligence Agency
CRISPR	clustered regularly interspaced short palindromic repeats
DoD	U.S. Department of Defense
EU	European Union
GDP	gross domestic product
GWP	gross world product
ICBM	intercontinental ballistic missile
IMF	International Monetary Fund
ISIS	Islamic State of Iraq and Syria, Islamic State of Iraq and the Sham (both abbreviated as ISIS)
NAFTA	North American Free Trade Agreement
NATO	North Atlantic Treaty Organization
OCO	Overseas Contingency Operations
OECD	Organisation for Economic Co-operation and Development
R&D	research and development
R2P	responsibility to protect

STEM	science, technology, engineering and mathematics
TPP	Trans Pacific Partnership
TTIP	Transatlantic Trade and Investment Partnership
UN	United Nations
USAID	U.S. Agency for International Development
WTO	World Trade Organization

References

"2014 Global R&D Funding Forecast," *R&D Magazine*, December 9, 2013. As of November 7, 2016:
http://www.rdmag.com/sites/rdmag.com/files/gff-2014-5_7%20875x10_0.pdf

Adams, John Quincy, speech on U.S. foreign policy, July 4, 1812.

Addison, Tony, and Finn Tarp, *Lessons for Japanese Foreign Aid from Research on Aid's Impact*, United Nations University—World Institute for Development Economics Research, WIDER Working Paper 2015/058, August 2015.

Albright, Madeleine, *Today*, NBC, February 19, 1998.

Alden, Edward, and Rebecca Strauss, *How America Stacks Up: Economic Competitiveness and U.S. Foreign Policy*, New York: Council on Foreign Relations, 2016.

al-Gharbi, Musa, "Why America Lacks Credibility in the Middle East," *Foreign Policy in Focus*, March 10, 2015. As of November 7, 2016:
http://fpif.org/america-lacks-credibility-middle-east/

Amano, Yukiua, *The Fukushima Daiichi Accident: Report by the Director General*, Vienna: International Atomic Energy Agency, 2015. As of November 2, 2016:
http://www-pub.iaea.org/books/IAEABooks/10962/The-Fukushima-Daiichi-Accident

American Association for the Advancement of Science, "Historical Trends in Federal R&D," web page, 2015. As of November 7, 2016:
https://www.aaas.org/page/historical-trends-federal-rd#Overview

American Society of Civil Engineers, "Report Card for America's Infrastructure," web page, 2013. As of November 7, 2016:
http://www.infrastructurereportcard.org

Anderson, Colonel John C., *Changing the Game: Human Security as Grand Strategy*, Carlisle, Pa.: United States Army War College Press, June 2014.

Arias, Elizabeth, "Changes in Life Expectancy by Race and Hispanic Origin in the United States, 2013–2014," Centers for Disease Control and Prevention, NCHS Data Brief, No. 244, April 2016.

Asia Society China File, "The Future of NGOs in China," May 14, 2015.

Autor, David H., David Dorn, and Gordon H. Hanson, "The China Syndrome: Local Labor Market Effects of Import Competition in the United States," *American Economic Review*, Vol. 103, No. 6, October 2013, pp. 2121–2168.

Autor, David H., David Dorn, Gordon Hanson, and Jae Song, "Trade Adjustment: Worker-Level Evidence," *Quarterly Journal of Economics*, Vol. 129, No. 4, 2014, pp. 1799–1860.

Avendano, Mauricio, and Ichiro Kawachi, "Why Do Americans Have Shorter Life Expectancy and Worse Health Than People in Other High-Income Countries?" *Annual Review of Public Health*, Vol. 35, 2014, pp. 307–325.

Bacevich, Andrew J., *America's War for the Greater Middle East: A Military History*, New York: Penguin Random House, 2016.

Baiocchi, Dave, and D. Steven Fox, *Surprise! From CEOs to Navy Seals: How a Select Group of Professionals Prepare for and Respond to the Unexpected*, Santa Monica, Calif.: RAND Corporation, RR-341-NRO, 2013. As of July 22, 2016: http://www.rand.org/pubs/research_reports/RR341.html

Baker, Jill, "Climate Expert Says China Committed to Battling Climate Change—Even If the U.S. Drops out of Paris Accord," Forbes.com, November 14, 2016. As of November 25, 2016:
http://www.forbes.com/sites/jillbaker/2016/11/14/climate-expert-says-china-committed-to-battling-climate-change-even-if-the-u-s-drops-out-of-paris-accord/#166e11767440

Balakrishnan, Ravi, Mai Dao, Juan Sole, and Jeremy Zook, "Lost Workers," International Monetary Fund, *Finance and Development*, Vol. 52, No. 3, September 2015, pp. 1–4.

Bamford, James, "Frozen Assets: The Newest Front in Global Espionage Is One of the Least Habitable Locales on Earth—the Arctic," *Foreign Policy*, May 11, 2015. As of November 25, 2016:
http://foreignpolicy.com/2015/05/11/frozen-assets-arctic-espionage-spying-new-cold-war-russia-canada/

Barrett, Anthony, *False Alarms, True Dangers? Current and Future Risks of Inadvertent U.S.-Russian Nuclear War*, Santa Monica, Calif.: RAND Corporation, PE-191-TSF, 2016. As of June 9, 2016:
http://www.rand.org/pubs/perspectives/PE191.html

Bass, Dina, "America's Unwanted Ivy Leaguers Are Flocking to India," Bloomberg, June 2, 2015. As of November 7, 2016:
http://www.bloomberg.com/news/articles/2015-06-02/chasing-the-american-dream-in-india

"Battle of the Declaration" (Chinese PLA recruiting video), Youku Tudou, website, 2016. As of November 7, 2016:
http://v.youku.com/v_show/id_XMTU1MjM3NTg5Mg==.html

Bernanke, Ben S., "Proceedings," The Defense Economy and American Prosperity, Washington, D.C., August 17, 2015. As of November 7, 2016:
http://www.brookings.edu/~/media/events/2015/08/17-defense-economy/20150817_defense_economy_transcript.pdf

Betts, Richard K., *Enemies of Intelligence: Knowledge and Power in American National Security*, Columbia, N.Y.: Columbia University Press, 2007.

Binnendijk, Hans, *Friends, Foes, and Future Directions: U.S. Partnerships in a Turbulent World*, Santa Monica, Calif.: RAND Corporation, RR-1210-RC, 2016. As of November 2, 2016:
http://www.rand.org/pubs/research_reports/RR1210.html

Biological and Toxin Weapons Convention Website, University of Bradford, undated. As of November 7, 2016:
http://www.opbw.org/

Bitzinger, Richard A., "China's Double-Digit Defense Growth: What It Means for a Peaceful Rise," *Foreign Affairs,* March 19, 2015.

Blackwell, Evan, *Prison and Jihad: A Threat Behind Bars*, NATO Association of Canada, October 10, 2015.

Blackwill, Robert D., and Jennifer M. Harris, "The Lost Art of Economic Statecraft," *Foreign Affairs*, Vol. 95, No. 2, March/April 2016.

Blackwill, Robert D., and Ashley J. Tellis, *Revising U.S. Grand Strategy Toward China*, Washington, D.C.: Council on Foreign Relations, March 2015.

Blain, Loz, "Do-It-Yourself CRISPR Genome Editing Kits Bring Genetic Engineering to Your Kitchen Bench," Gizmag.com, November 12, 2015. As of June 27, 2016:
http://www.gizmag.com/home-crispr-gene-editing-kit/40362/

Blanton, Thomas S., "The President's Daily Brief," Washington, D.C.: National Security Archive, April 12, 2004. As of November 2, 2016:
http://nsarchive.gwu.edu/NSAEBB/NSAEBB116/

Bliddal, Henrik, *Reforming Military Command Arrangements: The Case of the Rapid Deployment Joint Task Force*, Carlisle, Pa.: U.S. Army War College, 2011.

Bradford, Scott C., Paul L. E. Grieco, and Gary Hufbauer, "The Payoff to America From Global Integration," in C. Fred Bergsten, ed., *United States and the World Economy: Foreign Economic Policy for the Next Decade*, Washington, D.C.: Institute for International Economics, January 2005.

Brands, Hal, *What Good Is Grand Strategy? Power and Purpose in American Statecraft from Harry S. Truman to George W. Bush*, Ithaca, N.Y.: Cornell University Press, 2014.

———, *American Grand Strategy and the Liberal Order: Continuity, Change, and Options for the Future*, Santa Monica, Calif.: RAND Corporation, PE-209-OSD, 2016. As of November 06, 2016:
http://www.rand.org/pubs/perspectives/PE209.html

Breedlove, General Philip, *U.S. European Command Posture Statement 2016*, U.S. European Command, Stuttgart, Germany, February 25, 2016. As of June 24, 2016:
http://www.eucom.mil/media-library/article/35164/u-s-european-command-posture-statement-2016

Bremmer, Ian, *Superpower: Three Choices for America's Role in the World*, New York: Portfolio (Penguin Group), May 2015.

British Petroleum, *BP Statistical Review of World Energy 2014*, London, June 2014.

Brodie, Bernard, "The Anatomy of Deterrence," Research Memorandum, U.S. Air Force Project RAND, RM-2218, 1958.

Brooks, David, "Enter the Age of the Outsiders," *New York Times*, October 20, 2015.

Brynjolfsson, Erik, and Andrew McAfee, *The Second Machine Age*, New York: W.W. Norton, 2016.

Brzezinski, Zbigniew, *Strategic Vision: America and the Crisis of Global Power*, New York: Basic Books, 2012.

Buckley, Chris, "China Assails Textbooks Sowing 'Western Values,'" *New York Times*, January 31, 2015.

Bunn, M. Elaine, "Can Deterrence Be Tailored?" *Strategic Forum*, Institute for National Strategic Studies, National Defense University, No. 225, January 2007.

Burns, Nicholas, "Barack Obama Has Risked U.S. Credibility in the Middle East," *Financial Times*, March 22, 2016. As of November 7, 2016:
https://www.ft.com/content/3590f250-ef6b-11e5-9f20-c3a047354386

Burtless, Gary, "Unemployment and the 'Skills Mismatch' Story: Overblown and Unpersuasive," Brookings Institution, blog, July 29, 2014. As of November 7, 2016:
http://www.brookings.edu/research/opinions/2014/07/29-skills-mismatch-overblown-burtless

Callaway, Ewan, "'Minimal' Cell Raises Stakes in Race to Harness Synthetic Life," *Nature News*, Vol. 531, No. 7596, March 26, 2016.

Calmes, Jackie, "Hiring Hurdle: Finding Workers Who Can Pass a Drug Test," *New York Times*, May 17, 2016.

"Can She Fix It?" *The Economist*, April 23, 2016. As of November 7, 2016: http://www.economist.com/news/leaders/21697225-hillary-clinton-has-democratic-nomination-within-her-grasp-she-needs-bolder-ideas-what

Carlson, Benjamin, "7 Things You Can't Talk About in China," *Global Post*, June 3, 2013. As of November 10, 2016: http://www.pri.org/stories/2013-06-03/7-things-you-cant-talk-about-china

Carnevale, Anthony P., Tamara Jayasundera, and Artem Gulish, *Six Million Missing Jobs: The Lingering Pain of the Great Recession*, Washington, D.C.: Georgetown University Center on Education and the Workforce, 2015. As of November 7, 2016: https://cew.georgetown.edu/wp-content/uploads/Six-Million-Missing-Jobs.pdf

Carnevale, Anthony P., Nicole Smith, and Jeff Strohl, *Recovery: Job Growth and Education Requirements Through 2020*, Washington, D.C.: Georgetown Public Policy Institute, Georgetown University, June 2013.

Carothers, Thomas, Lina Khatib, Marwan Muasher, Douglas H. Paal, and Andrew S. Weiss, "Is the World Falling Apart?" Carnegie Endowment for International Peace, August 14, 2014. As of December 16, 2015: http://carnegieendowment.org/2014/08/14/is-world-falling-apart

Carter, Ashton, U.S. Secretary of Defense, *Submitted Statement—Senate Armed Services Committee (Budget Request)*, Washington, D.C., March 3, 2015.

Carter, Jimmy, *The State of the Union Address Delivered Before a Joint Session of the Congress,* January 23, 1980. Online by Gerhard Peters and John T. Woolley, The American Presidency Project.As of December 15, 2016: http://www.presidency.ucsb.edu/ws/?pid=33079

Case, Anne, and Angus Deaton, "Rising Morbidity and Mortality in Midlife Among White Non-Hispanic Americans in the 21st Century," *Proceedings of the National Academy of Sciences*, Vol. 112, No. 49, December 8, 2015, pp. 15078–15083.

CBO—*See* Congressional Budget Office.

Center on Global Energy Policy, "Summary of the Roundtable on the Resilience of the Liquid Fuel Supply Chain in the New York City Area," Columbia University, May 28, 2014.

Chang, Welton, "Five Psychological Biases That Plague Policymakers and Military Leaders," *The Policy*, February 8, 2016.

Chase, Michael S., "China's Transition to a More Credible Nuclear Deterrent: Implications and Challenges for the United States," *Asia Policy*, July 2013, pp. 69–101.

Chestney, Nina, and Alister Doyle, "France, U.N. Tell Trump Action on Climate Change Unstoppable," Reuters, November 15, 2016. As of November 17, 2016: http://www.reuters.com/article/us-climatechange-accord-idUSKBN13A12Z

"China Wants Quick Close on Regional Trade Agreement for Asia," Bloomberg News, November 24, 2016. As of November 26, 2016: http://www.bloomberg.com/news/articles/2016-11-24/china-wants-quick-close-on-regional-trade-pact-after-tpp-dashed

Chinese Communist Party Central Committee, "Communiqué on the Current State of the Ideological Sphere," Asia Society China File translation, November 8, 2013.

Chollet, Derek, *The Long Game: How Obama Defied Washington and Redefined America's Role in the World*, New York: Public Affairs, 2016.

Christensen, Thomas J., "The Meaning of the Nuclear Evolution: China's Strategic Modernization and U.S.-China Security Relations," *Journal of Strategic Studies*, August 2012, pp. 447–487.

Clapper, James R., Director of National Intelligence, "Worldwide Cyber Threats," statement to House Permanent Select Committee on Intelligence, September 10, 2015.

———, "Worldwide Threat Assessment of the U.S. Intelligence Community," statement for the record to the Senate Select Committee on Intelligence, February 9, 2016.

Cliff, Roger, *China's Military Power: Assessing Current and Future Capabilities*, Cambridge, UK: Cambridge University Press, September 2015.

Clinton, William J., "Remarks on the Economic Program in Santa Monica, California," February 21, 1993.

CNN/ORC International, "CNN/ORC International Poll," December 6, 2015. As of June 22, 2016: http://i2.cdn.turner.com/cnn/2015/images/12/06/rel12d.-.obama.isis.pdf

———, "CNN/ORC International Poll," May 5, 2016. As of June 22, 2016: http://i2.cdn.turner.com/cnn/2016/images/05/05/rel6c.-.isis.pdf

Congressional Budget Office, *Long-Term Implications of the 2016 Future Years Defense Program*, Washington, D.C., CBO 51050, January 14, 2016. As of June 14, 2016: http://www.cbo.gov/publication/51050

Consumer and Community Development Research Section of the Federal Reserve Board's Division of Consumer and Community Affairs, *Report on the Economic Well-Being of U.S. Households in 2015*, Washington, D.C.: Board of Governors of the Federal Reserve System, May 2016.

Consumer Financial Protection Bureau, *Student Loan Servicing: Analysis of Public Input and Recommendations for Reform*, Washington, D.C., September 2015. As of November 7, 2016:
http://files.consumerfinance.gov/f/201509_cfpb_student-loan-servicing-report.pdf

Cordesman, Anthony H., "Strategy, Grand Strategy, and the Emperor's New Clothes," *Commentary*, Washington, D.C.: Center for Strategic and International Studies, March 17, 2015.

Costantino, Maureen, and Leigh Angres, "The Federal Budget in 2015," fact sheet, Congressional Budget Office, Washington, D.C., January 2016. As of June 7, 2016:
https://www.cbo.gov/sites/default/files/114th-congress-2015-2016/graphic/51110-budget1overall.pdf

Courtney, William, and Michael Haltzel, "Putin is Trying to Bully Europe into Dropping Sanctions," RAND blog, June 11, 2016. As of June 23, 2016:
http://www.rand.org/blog/2016/06/putin-is-trying-to-bully-europe-into-dropping-sanctions.html

Cowan, Tyler, "Skill Mismatch Unemployment is Real and Significant," MarginalRevolution, website, November 21, 2015. As of November 7, 2016:
http://marginalrevolution.com/marginalrevolution/2015/11/skill-mismatch-unemployment-is-real-and-significant.html

Crane, Keith, Shanthi Natraj, Patrick B. Johnston, and Gursel Rafig oglu Aliyev, *Russia's Medium-Term Economic Prospects*, Santa Monica, Calif.: RAND Corporation, RR-1468-RC, 2016. As of November 2, 2016:
http://www.rand.org/pubs/research_reports/RR1468.html

Crocker, Chester, "The Strategic Dilemma of a World Adrift," *Survival*, Vol. 57, No. 1, February–March 2015.

Cronin, Audrey Kurth, "How al-Qaida Ends: The Decline and Demise of Terrorist Groups," *International Security*, Vol. 31, No. 1, Summer 2006, pp. 7–48.

———, *How Terrorism Ends: Understanding the Decline and Demise of Terrorist Campaigns*, Princeton, N.J.: Princeton University Press, 2009.

Cunningham, Fiona S., and M. Taylor Fravel, "Assuring Assured Retaliation: China's Nuclear Posture and U.S.-China Strategic Stability," *International Security*, Fall 2015, pp. 7–50.

Dabla-Norris, Era, Kalpana Kochhar, Nujin Suphaphiphat, Frantisek Ricka, and Evridiki Tsounta, *Causes and Consequences of Income Inequality: A Global Perspective*, Washington, D.C.: International Monetary Fund, June 2015.

Davis, Major John R., Jr., "Continued Evolution of Hybrid Threats: The Russian Hybrid Threat Construct and the Need for Innovation," *Three Swords Magazine*, Vol. 28, 2015, pp. 19–25. As of November 7, 2016:
http://www.jwc.nato.int/images/stories/threeswords/
CONTINUED_EVOLUTION_OF_HYBRID_THREATS.pdf

Davis, Lynn E., Michael J. McNerney, and Daniel Byman, "Armed Drone Myth 1: They Will Transform How War is Waged Globally," The RAND Blog, February 17, 2015. As of November 7, 2016:
http://www.rand.org/blog/2015/02/
armed-drone-myth-1-they-will-transform-how-war-is-waged.html

DeConto, Robert M., and David Pollard, "Contribution of Antarctica to Past and Future Sea-Level Rise," *Nature*, Vol. 531, No. 7596, March 31, 2016, pp. 591–597.

Demirjian, Karoun, and Juliet Eilperin, "Congress Overrides Obama's Veto of 9/11 Bill," *Washington Post*, September 28, 2016.

Denning, Steve, "It's Official! The End of Competitive Advantage," Forbes.com, June 2, 2013. As of December 16, 2015:
http://www.forbes.com/sites/stevedenning/2013/06/02/
its-official-the-end-of-competitive-advantage/

Desai, Mihir A., C. Fritz Foley, and Kristin Forbes, "Financial Constraints and Growth: Multinational and Local Firm Responses to Currency Depreciations," *Review of Financial Studies*, Vol. 21, No. 6, November 2008, pp. 2857–2888.

Deudney, Daniel, and G. John Ikenberry, *Democratic Internationalism: An American Strategy for a Post-Exceptionalist Era*, New York: Council on Foreign Relations, working paper, November 2012.

Dewar, James A., *Assumption-Based Planning—A Tool for Reducing Avoidable Surprises*, Cambridge: Cambridge University Press, 2002.

Diamond, Larry, "Facing Up to the Democracy Recession," *Journal of Democracy*, Vol. 26, No. 1, January 2015, pp. 141–155.

Dobbins, James, Richard H. Solomon, Michael S. Chase, Ryan Henry, F. Stephen Larrabee, Robert J. Lempert, Andrew Liepman, Jeffrey Martini, David Ochmanek, and Howard J. Shatz, *Choices for America in a Turbulent World : Strategic Rethink*, Santa Monica, Calif.: RAND Corporation, RR-1114-RC, 2015. As of October 28, 2016:
http://www.rand.org/pubs/research_reports/RR1114.html

Dobbs, Richard, Anu Madgavkar, Dominic Barton, Eric Labaye, James Manyika, Charles Roxburgh, Susan Lund, and Siddarth Madhav, *The World at Work: Jobs, Pay and Skills for 3.5 Billion People*, McKinsey Global Institute, June 2012. As of November 7, 2016:
http://www.mckinsey.com/global-themes/employment-and-growth/
the-world-at-work

Doherty, Patrick C., *A New U.S. Grand Strategy*, Sacramento, Calif.: New America Foundation, January 9, 2013.

Drouet, Laurent, Valentina Bosetti, and Massimo Tavoni, "Selection of Climate Policies Under the Uncertainties in the Fifth Assessment Report of the IPCC," *Nature Climate Change*, Vol. 5, February 2015, pp. 937–940.

Duran, Charles, "Power Cycle Theory, the Shifting Tides of History, and Statecraft: Interpreting China's Rise," *SAIS Europe Journal of Global Affairs*, April 1, 2012.

Dwyer, Paula, "A Basic Income Should Be the Next Big Thing," *Bloomberg View*, May 2, 2016.

Economic Innovation Group, "The 2016 Distressed Communities Index: An Analysis of Community Well-Being Across the United States," February 25, 2016. As of November 16, 2016:
http://eig.org/wp-content/uploads/2016/02/2016-Distressed-Communities-Index-Report.pdf

Edelstein, David M., and Ronald R. Krebs, "Delusions of Grand Strategy," *Foreign Affairs*, December 2015.

Egel, Daniel, Adam R. Grissom, John P. Godges, Jennifer Kavanagh, and Howard J. Shatz, *Estimating the Value of Overseas Security Commitments*, Santa Monica, Calif.: RAND Corporation, RR-518-AF, 2016. As of October 27, 2016:
http://www.rand.org/pubs/research_reports/RR518.html

Environmental Protection Agency, "Carbon Dioxide Capture and Sequestration," web page, undated. As of November 17, 2016:
https://www.epa.gov/climatechange/carbon-dioxide-capture-and-sequestration-overview

Erickson, Andrew S., "PLA Recruiting Ad: The Rap Video—Annotated Translations by Dr. Kevin Jensen and Professor Mark Metcalf," Andrew S. Erickson, China Analysis from Original Sources, blog, May 11, 2016. As of November 7, 2016:
http://www.andrewerickson.com/2016/05/pla-recruiting-ad-the-rap-video-annotated-translation-by-dr-kevin-jensen-prof-mark-metcalf/

Fairlie, Robert W., Arnobio Morelix, E. J. Reedy, and Joshua Russell, *The Kauffman Index: Startup Activity: National Trends*, Kansas City, Mo.: Ewing Marion Kauffman Foundation, 2015. As of October 2016:
http://www.kauffman.org/~/media/kauffman_org/research%20reports%20and%20covers/2015/05/kauffman_index_startup_activity_national_trends_2015.pdf

Fallows, James, "Japan: Let Them Defend Themselves," *The Atlantic*, April 1989.

Federation of American Scientists, "Intelligence Budget Data," web page, 2016. As of June 27, 2016:
https://fas.org/irp/budget/

Fields, Gary, and John R. Emshwiller, "As Arrest Records Rise, Americans Find Consequences Can Last a Lifetime: Even When Charges Are Dropped, a Lingering Arrest Record Can Ruin Chances of a Job," *Wall Street Journal*, August 18, 2014.

Frank, Barney, "It's Time to Rearm Germany and Japan," *Politico*, October 21, 2015. As of December 16, 2015:
http://www.politico.com/magazine/story/2015/10/its-time-to-rearm-germany-and-japan-213279

Frankel, Jeffrey A., and David Romer, "Does Trade Cause Growth?" *American Economic Review*, Vol. 89, No. 3, June 1999, pp. 379–399.

Fravel, M. Taylor, and Evan S. Medeiros, "China's Search for Assured Retaliation: The Evolution of Chinese Nuclear Strategy and Force Structure," *International Security*, Fall 2010, pp. 48–87.

Freedman, Andrew, "The Last Time CO2 Was This High, Humans Didn't Exist," *Climate Central*, May 3, 2013. As of June 17, 2015:
http://www.climatecentral.org/news/the-last-time-co2-was-this-high-humans-didnt-exist-15938

Freedman, Lawrence, *Strategy: A History*, New York: Oxford University Press, 2013.

Freedom House, *Freedom in the World 2016: Anxious Dictators, Wavering Democracies*, Washington, D.C., 2016a. As of November 7, 2016:
https://freedomhouse.org/article/new-report-freedom-world-2016-anxious-dictators-wavering-democracies

———, "Freedom in the World: Country Ratings 1973–2016," 2016b.

———, "Maps and Graphics," web page, 2016c. As of November 7, 2016:
https://freedomhouse.org/report/freedom-world-2016/maps-graphics

Freeman, Chas W., Jr., "America in the New World Disorder," remarks at the Watson Institute for International and Public Affairs, Brown University, February 11, 2016.

Freier, Nathan, *Known Unknowns: Unconventional 'Strategic Shocks' in Defense Strategy Development*, Carlisle, Pa.: U.S. Army War College, November 2008.

Friedman, Thomas L., *The Lexus and the Olive Tree: Understanding Globalization*, New York: Farrar, Straus and Giroux, 1999.

———, "What If?" *New York Times*, January 20, 2016.

Fuerth, Leon, and Evan M. H. Farber, *Anticipatory Governance, Practical Upgrades: Equipping the Executive Branch to Cope with Increasing Speed and Complexity of Major Challenges*, Washington, D.C.: The Project on Forward Engagement, October 2012.

Fukuyama, Francis, "American Political Decay or Renewal? The Meaning of the 2016 Election," *Foreign Affairs*, Vol. 95, No. 4, July/August 2016.

Fung, Brian, "Obama Called the Sony Hack an Act of 'Cyber Vandalism.' He's Right." *Washington Post*, December 22, 2014. As of November 8, 2016: https://www.washingtonpost.com/news/the-switch/wp/2014/12/22/obama-called-the-sony-hack-an-act-of-cyber-vandalism-hes-right/

Gaddis, John Lewis, *Strategies of Containment: A Critical Appraisal of American National Security Policy During the Cold War*, New York: Oxford University Press, 1982.

———, *The Cold War: A New History*, New York: Penguin Press, 2005.

———, "What Is Grand Strategy?" Heyden Distinguished Lecture, Duke University, February 26, 2009.

Gaddy, Clifford, and Michael O'Hanlon, "Toward a 'Reaganov' Russia: Russian Security Policy After Putin," *Washington Quarterly*, July 2015.

Galeotti, Mark, and Andrew S. Bowen, "Putin's Empire of the Mind," *Foreign Policy*, May/June 2014, pp. 16–19.

Garrett, Laurie, *CRISPR: Transformative and Troubling*, Council on Foreign Relations, April 13, 2016.

Gartzke, Erik, Jon Lindsay, and Michael Nacht, "Cross-Domain Deterrence: Strategy in an Era of Complexity," International Studies Association Annual Meeting, Toronto, 2014.

Gellman, Barton, and Greg Miller, "'Black Budget' Summary Details U.S. Spy Network's Successes, Failures and Objectives," *Washington Post*, August 29, 2013. As of November 10, 2016: https://www.washingtonpost.com/world/national-security/black-budget-summary-details-us-spy-networks-successes-failures-and-objectives/2013/08/29/7e57bb78-10ab-11e3-8cdd-bcdc09410972_story.html?utm_term=.c4c0f76a179b

Gerstein, Daniel M., "How Genetic Editing Became a National Security Threat," *Bulletin of the Atomic Scientists*, April 25, 2016.

Gilbert, Daniel, "If Only Gay Sex Caused Global Warming," *Los Angeles Times*, July 2, 2006.

"Global Warming," CBS News/New York Times Poll, November 18–22, 2015. As of November 7, 2016: https://www.scribd.com/doc/291657688/CBS-News-Global-Warming-Poll-Toplines

Goforth, Sean,"Brazil's Erratic Behavior," *The National Interest*, April 11, 2012. As of June 23, 2016: http://nationalinterest.org/commentary/brazils-erratic-behavior-6767

Goldin, Ian, and Mike Mariathasan, *The Butterfly Defect: How Globalization Creates Systemic Risks, and What to Do About It*, Princeton, N.J.: Princeton University Press, May 11, 2014.

Gompert, David C., Hans Binnendijk, and Bonny Lin, *Blinders, Blunders, and Wars: What America and China Can Learn*, Santa Monica, Calif.: RAND Corporation, RR-768-RC, 2014. As of May 3, 2016:
http://www.rand.org/pubs/research_reports/RR768.html

Gompert, David C., Astrid Cevallos, and Cristina L. Garafola, *War with China: Thinking Through the Unthinkable*, Santa Monica, Calif.: RAND Corporation, RR-1140-A, 2016. As of August 22, 2016:
http://www.rand.org/pubs/research_reports/RR1140.html

Gordon, Philip H., *Winning the Right War: The Path to Security for America and the World*, New York: Henry Holt, 2007.

Gosselin, Peter, and Mike Dorning, "After Doubts, Economists Find China Kills U.S. Factory Jobs," Bloomberg, June 18, 2015.

Gould, Elise, "Still No Sign of a Skills Mismatch—Unemployment is Elevated Across the Board," Economic Policy Institute, blog, January 13, 2015. As of November 8, 2016:
http://www.epi.org/blog/
still-no-sign-of-a-skills-mismatch-unemployment-is-elevated-across-the-board/

Gunther McGrath, Rita, and Alex Gourlay, *The End of Competitive Advantage: How to Keep Your Strategy Moving as Fast as Your Business*, Boston, Mass.: Harvard Business Publishing, June 4, 2013.

Hagel, Charles T., *Quadrennial Defense Review*, Washington, D.C.: U.S. Department of Defense, March 2014.

Hall, Keith, *The 2015 Long-Term Budget Outlook*, Washington, D.C.: Congressional Budget Office, June 16, 2015.

Hammes, T. X., "Will Technological Convergence Reverse Globalization?" *Strategic Forum*, Institute for National Strategic Studies, National Defense University, SF No. 297, July 2016.

Hart, B. H. Liddell, *Strategy*, Second Revised Edition, New York: Meridian Books, 1991.

Heginbotham, Eric, Michael Nixon, Forrest E. Morgan, Jacob Heim, Jeff Hagen, Sheng Li, Jeffrey Engstrom, Martin C. Libicki, Paul DeLuca, David A. Shlapak, David R. Frelinger, Burgess Laird, Kyle Brady, and Lyle J. Morris, *The U.S.-China Military Scorecard: Forces, Geography, and the Evolving Balance of Power, 1996–2017*, Santa Monica, Calif.: RAND Corporation, RR-392-AF, 2015. As of October 28, 2016:
http://www.rand.org/pubs/research_reports/RR392.html

Hill, Charles, *Grand Strategies: Literature, Statecraft, and World Order*, New Haven, Conn.: Yale University Press, 2010.

Hill, Fiona, and Clifford G. Gaddy, *Mr. Putin, Operative in the Kremlin*, Washington, D.C.: Brookings Institution Press, 2015.

Hoehn, Andrew R., and Sarah Harting, *Risking NATO: Testing the Limits of the Alliance in Afghanistan*, Santa Monica, Calif.: RAND Corporation, MG-974-AF, 2010. As of December 16, 2015:
http://www.rand.org/pubs/monographs/MG974.html

Holmes, James R., "Japan: Joining the Nuclear Weapons Club? It Could," *The Diplomat*, October 22, 2012. As of November 8, 2016:
http://thediplomat.com/2012/10/japan-joining-the-nuclear-weapons-club-it-could/

Holyk, Gregory, "Foreign Policy in the 2016 Presidential Primaries Based on the Exit Polls," The Chicago Council on Global Affairs, blog, April 7, 2016. As of November 8, 2016:
https://www.thechicagocouncil.org/blog/running-numbers/foreign-policy-2016-presidential-primaries-based-exit-polls

Hooker, R. D., Jr., *The Grand Strategy of the United States*, Washington, D.C.: National Defense University, October 2014.

Horovitz, David, "George Shultz: In a World Awash in Change, Follow the Reagan Playbook," *Times of Israel*, February 21, 2016. As of November 8, 2016:
http://www.timesofisrael.com/george-shultz-in-a-world-awash-in-change-follow-the-reagan-playbook/

Hudson, John, "Senators Slam NATO 'Free-Riders' in Closed-Door Meeting with Secretary General," *Foreign Policy*, April 6, 2016. As of November 8, 2016:
http://foreignpolicy.com/2016/04/06/senators-slam-nato-free-riders-in-closed-door-meeting-with-secretary-general/

Human Security Centre, *Human Security Report 2005: War and Peace in the 21st Century,* New York: Oxford University Press, 2005.

Hurd, Michael D., Paco Martorell, Adeline Delavande, Kathleen J. Mullen, and Kenneth M. Langa, "Monetary Costs of Dementia in the United States," *New England Journal of Medicine*, Vol. 368, No. 14, April 4, 2013, pp. 1326–1334.

Hutchison, Clyde A. III, Ray-Yuan Chuang, Vladimir N. Noskov, Nacyra Assad-Garcia, Thomas J. Deerinck, Mark H. Ellisman, John Gill, Krishna Kannan, Bogumil J. Karas, Li Ma, James F. Pelletier, Zhi-Qing Qi, R. Alexander Richter, Elizabeth A. Strychalski, Lijie Sun, Yo Suzuki, Billyana Tsvetanova, Kim S. Wise, Hamilton O. Smith, John I. Glass, Chuck Merryman, Daniel G. Gibson, and J. Craig Venter, "Design and Synthesis of a Minimal Bacterial Genome," *Science*, Vol. 351, No. 6280, March 25, 2016.

"Ideas and Consequences," *The Atlantic*, October 2007. As of November 2, 2016: http://www.theatlantic.com/magazine/archive/2007/10/ideas-and-consequences/306193/

IFC Advisory Services, *Closing the Credit Gap for Formal and Informal Micro, Small, and Medium Enterprises*, Washington, D.C.: International Finance Corporation, 2013. As of November 17, 2016: https://www.ifc.org/wps/wcm/connect/4d6e6400416896c09494b79e78015671/Closing+the+Credit+Gap+Report-FinalLatest.pdf?MOD=AJPERES

Ikenberry, G. John, *Liberal Leviathan: The Origins, Crisis, and Transformation of the American World Order*, Princeton, N.J.: Princeton University Press, 2011.

Inbar, Efraim, and Eitan Sharim, *Mowing the Grass in Gaza*, Begin-Sadat Center for Strategic Studies, Perspective Paper No. 255, July 20, 2014. As of August 22, 2014: http://besacenter.org/perspectives-papers/mowing-grass-gaza

Inglehart, Ronald F., and Pippa Norris, *Trump, Brexit, and the Rise of Populism: Economic Have-Nots and Cultural Backlash*, Cambridge, Mass.: Harvard University Kennedy School of Government Faculty Working Paper, RWP16-026, August 2016. As of November 18, 2016: https://research.hks.harvard.edu/publications/getFile.aspx?Id=1401

"Inside the 2013 U.S. Intelligence 'Black Budget,'" *Washington Post*, August 29, 2013. As of October 31, 2016: https://www.washingtonpost.com/apps/g/page/national/inside-the-2013-us-intelligence-black-budget/420/

Intelligence Advanced Research Projects Activity, "Decoding the Future for National Security," web page, undated-a. As of March 28, 2016: https://www.iarpa.gov/index.php/newsroom/iarpa-in-the-news/2015/646-decoding-the-future-for-national-security

———, "Office for Anticipating Surprise and the 'Super Forecasters,'" web page, undated-b. As of March 28, 2016: https://www.iarpa.gov/index.php/newsroom/iarpa-in-the-news/2015/555-office-for-anticipating-surprise-and-the-super-forecasters

International Monetary Fund, "The Macroeconomic Effects of Public Investment," in *World Economic Outlook, Legacies, Clouds, Uncertainties*, Washington, D.C., October 2014.

———,"Is It Time for an Infrastructure Push?" IMF World Economic Outlook Database, October 2016a.

———, World Economic Outlook Database, October 2016b.

International Rivers, "Map of Climate Change Hotspots," (CC BY-NC-SA 3.0 US), undated. As of November 10, 2016: https://www.internationalrivers.org/resources/map-of-climate-change-hotspots-3502

"Iran's Leader Tells IRGC to Guard Against Western Influence," Payvand Iran News, September 17, 2015.

Irwin, Neil, "What Was the Greatest Era for Innovation? A Brief Guided Tour," *New York Times*, May 13, 2016.

Iyengar, Radha, and Rebecca Friedman Lissner, "Iran, Terrorism, and Nonproliferation After the Nuclear Deal," *War on the Rocks* blog, January 28, 2016. As of December 16, 2016:
http://warontherocks.com/2016/01/iran-terrorism-and-nonproliferation-after-the-nuclear-deal/

Jain, Ash, *Like-Minded and Capable Democracies: A New Framework for Advancing a Liberal World Order*, New York: Council on Foreign Relations, January 2013.

Jaruzelski, Barry, Kevin Schwartz, and Volker Staack, "The 2015 Global Innovation 1000: Innovation's New World Order," *Strategy and Business*, No. 81, Winter, October 27, 2015.

Jentleson, Bruce W., "Strategic Recalibration: Framework for a 21st Century National Security Strategy," *Washington Quarterly*, Spring 2014.

Jones, Bruce, *Still Ours to Lead: America, Rising Powers, and the Tension Between Rivalry and Restraint*, Washington, D.C.: Brookings Institution Press, 2014.

Jones, Charles I., *The Facts of Economic Growth*, Cambridge, Mass.: National Bureau of Economic Research, NBER Working Paper No. 21142, 2015.

Jones, Gregory S., "ISIS and Dirty Bombs," *The Cipher*, June 3, 2016.

Jones, Seth G., Andrew Liepman, and Nathan Chandler, *Counterterrorism and Counterinsurgency in Somalia: Assessing the Campaign Against Al Shabaab*, Santa Monica, Calif.: RAND Corporation, RR-1539-OSD, 2016. As of September 13, 2016:
http://www.rand.org/pubs/research_reports/RR1539.html

Joyner, James, "Europe's Free Ride on the American-Defense Gravy Train," National Review Online, July 13, 2014. As of December 16, 2015:
http://nationalinterest.org/feature/europes-free-ride-the-american-defense-gravy-train-10864

Kagan, Robert, *The World America Made*, New York: Knopf, 2012.

———, "Superpowers Don't Get to Retire," *New Republic*, May 26, 2014.

Kane, Angela, UN High Representative for Disarmament Affairs, video message to the Biological Weapons Convention: 40th Anniversary Event, Geneva, Switzerland, March 30, 2015. As of November 8, 2016:
https://ar-ar.facebook.com/odaunrec/videos/687652498023709/

Kaplan, Rebecca, *CBS This Morning*, May 18, 2015.

Kaplan, Sarah, and Jackie VanderBrug, "The Rise of Gender Capitalism," *Stanford Social Innovation Review*, Fall 2014. As of June 23, 2016:
http://ssir.org/articles/entry/the_rise_of_gender_capitalism

Kapusta, Phillip, "The Challenge of Operating in the Gray Zone," *Special Warfare*, Vol. 28, No. 4, October–December 2015, pp. 16–25.

Katz, Lawrence F., and Alan B. Krueger, *The Rise and Nature of Alternative Work Arrangements in the United States, 1995–2015*, Harvard University, Princeton University, and the National Bureau of Economic Research, March 29, 2016. As of November 8, 2016:
https://krueger.princeton.edu/sites/default/files/akrueger/files/katz_krueger_cws_-_march_29_20165.pdf

Kavanagh, Jennifer, *U.S. Security-Related Agreements in Force Since 1955: Introducing a New Database*, Santa Monica, Calif.: RAND Corporation, RR-736-AF, 2014. As of October 27, 2016:
http://www.rand.org/pubs/research_reports/RR736.html

Kelly, Nora, "The Senate Goes Home Without Funding Zika," *The Atlantic*, April 29, 2016.

Kelly, Terrence, James Dobbins, David A. Shlapak, David C. Gompert, Eric Heginbotham, Peter Chalk, and Lloyd Thrall, *The U.S. Army in Asia, 2030–2040*, Santa Monica, Calif.: RAND Corporation, RR-474-A, 2014. As of November 3, 2016:
http://www.rand.org/pubs/research_reports/RR474.html

Kennedy, John F., and Charles De Gaulle, "Memorandum of Conversation," Paris, May 31, 1961, posted online in U.S. Department of State, Office of the Historian, *Foreign Relations of the United States, 1961–1963*, Vol. XIV, *Berlin Crisis, 1961–1962*. As of June 12, 2016:
https://history.state.gov/historicaldocuments/frus1961-63v14/d30

Kerry, John F., "Remarks at the Strong Cities Network International Visitors Leadership Program for Municipal Leaders and Countering Violent Extremism Experts Event," Washington, D.C.: U.S. Department of State, March 1, 2016a.

———, "Remarks on Daesh and Genocide," U.S. Department of State, Washington, D.C., March 17, 2016b.

Kissinger, Henry, *Years of Renewal*, New York: Simon and Schuster, 1999.

———, *Does America Need a Foreign Policy?* New York: Simon and Schuster, 2001.

———, *World Order*, New York: Penguin Press, 2014.

———, "Global Challenges and the U.S. National Security Strategy," Testimony before the Senate Armed Services Committee, January 29, 2015. As of December 16, 2015:
http://www.armed-services.senate.gov/imo/media/doc/15-05%20-%201-29-15.pdf

Koppel, Ted, *Lights Out: A Cyberattack, a Nation Unprepared, Surviving the Aftermath*, New York: Crown Publishers, 2015.

Kosanovich, Karen, and Elini T. Sherman, *Trends in Long-Term Unemployment*, Washington, D.C.: Bureau of Labor Statistics, March 2015. As of November 15, 2016:
http://www.bls.gov/spotlight/2015/long-term-unemployment/

Kramer, Andrew E., "Russia Stakes New Claim to Expanse in the Arctic," *New York Times*, August 4, 2015. As of November 18, 2016:
http://www.nytimes.com/2015/08/05/world/europe/kremlin-stakes-claim-to-arctic-expanse-and-its-resources.html

Krepinevich, Andrew F., and Barry D. Watts, *Regaining Strategic Competence*, Washington, D.C.: Center for Strategic and Budgetary Assessments, 2009.

Krugman, Paul, "CSI: Trade Deficit," *New York Times*, April 24, 2006. As of September 14, 2016:
http://query.nytimes.com/gst/fullpage.html?res=9807E4DE123FF937A15757C0A9609C8B63

Lagarde, Christine, "We Need Forceful Policies to Avoid the Low-Growth Trap," IMF Direct blog, International Monetary Fund, September 1, 2016. As of November 25, 2016:
https://blog-imfdirect.imf.org/2016/09/01/we-need-forceful-policies-to-avoid-the-low-growth-trap/

Lake, Anthony, "Confronting Backlash States," *Foreign Affairs*, Vol. 73, No. 2, March/April 1994, pp. 45–55.

Landers, Peter, "The Man Who Predicted the Tsunami," *Wall Street Journal*, April 9, 2011. As of November 8, 2016:
http://www.wsj.com/articles/SB10001424052748704101604576248722573203608

Lane, Jason, "U.S. Losing Its Dominance in Global Higher Education Market," *The Conversation*, October 14, 2015.

Langa, Kenneth M., "Is the Risk of Alzheimer's Disease and Dementia Declining?" *Alzheimer's Research and Therapy*, Vol. 7, No. 1, March 26, 2015, pp. 1–4. As of November 8, 2016:
https://www.ncbi.nlm.nih.gov/pmc/articles/PMC4374373/

"Larry Summers Reflects," *Harvard Magazine*, April 15, 2016. As of May 29, 2016:
http://harvardmagazine.com/2016/04/larry-summers-reflects

Lawder, David, "Lew Says Excess Capacity 'Corrosive' for China Growth," Reuters, June 5, 2016.

Lempert, Robert J., "Can Scenarios Help Policymakers Be Both Bold and Careful?" in Francis Fukuyama, ed., *Blindside: How to Anticipate Forcing Events and Wild Cards in Global Politics*, Washington, D.C.: Brookings Institution Press, 2007.

———, "Scenarios That Illuminate Vulnerabilities and Robust Responses," *Climatic Change*, Vol. 117, No. 4, 2013, pp. 627–646.

Lempert, Robert J., Steven W. Popper, and Steven C. Bankes, *Shaping the Next One Hundred Years: New Methods for Quantitative, Long-Term Policy Analysis*, Santa Monica, Calif.: RAND Corporation, MR-1626-RPC, 2003. As of November 2, 2016:
http://www.rand.org/pubs/monograph_reports/MR1626.html

Lempert, Robert J., Drake Warren, Ryan Henry, Robert W. Button, Jonathan Klenk, and Kate Giglio, *Defense Resource Planning Under Uncertainty: An Application of Robust Decision Making to Munitions Mix Planning*, Santa Monica, Calif.: RAND Corporation, RR-1112-OSD, 2016. As of December 5, 2016:
http://www.rand.org/pubs/research_reports/RR1112.html

"Letters to the Editor: The Perils of Prediction, June 2nd," *The Economist*, July 15, 2007. As of November 8, 2016:
http://www.economist.com/blogs/theinbox/2007/07/the_perils_of_prediction_june

Lewis, Jeffrey, "How Long for Japan to Build a Deterrent?" *Arms Control Wonk*, blog, December 28, 2006. As of November 8, 2016:
http://www.armscontrolwonk.com/archive/201339/japans-nuclear-status/

Libicki, Martin C., Lillian Ablon, and Tim Webb, *The Defender's Dilemma: Charting a Course Toward Cybersecurity*, Santa Monica, Calif.: RAND Corporation, RR-1024-JNI, 2015. As of November 2, 2016:
http://www.rand.org/pubs/research_reports/RR1024.html

Libicki, Martin C., Howard J. Shatz, and Julie E. Taylor, *Global Demographic Change and Its Implications for Military Power*, Santa Monica, Calif.: RAND Corporation, MG-1091-AF, 2011. As of October 31, 2016:
http://www.rand.org/pubs/monographs/MG1091.html

Lilla, Ryan, "The Consequentialist," *New Yorker*, May 2, 2011.

Luttwak, Edward N., *Strategy: The Logic of War and Peace*, Cambridge, Mass.: Harvard University Press, 1987.

Lyons, Kate, and Gordon Darroch, "Frexit, Nexit or Oexit? Who Will Be Next to Leave the EU?" *The Guardian*, June 27, 2016.

MacDonald, Paul K., and Joseph M. Parent, "Graceful Decline? The Surprising Success of Great Power Retrenchment," *International Security*, Vol. 35, No. 4, Spring 2011, pp. 7–44.

Mahbubani, Kishore, *The New Asian Hemisphere: The Irresistible Shift of Global Power to the East*, New York: Public Affairs, 2009.

Marshall, Monty G., and Benjamin R. Cole, *Global Report 2014: Conflict, Governance, and State Fragility*, Vienna, Va.: Center for Systemic Peace, 2014. As of November 8, 2016:
http://www.systemicpeace.org/vlibrary/GlobalReport2014.pdf

Mastroe, Caitlin, and Susan Szmania, *Surveying CVE Metrics in Prevention, Disengagement and De-Radicalization Programs*, College Park, Md.: National Consortium for the Study of Terrorism and Responses to Terrorism, Report to the Office of University Programs, Science and Technology Directorate, Department of Homeland Security, 2016. As of November 8, 2016: http://www.start.umd.edu/pubs/START_SurveyingCVEMetrics_March2016.pdf

Mathews, Jessica, "What Foreign Policy for the US?" *New York Review of Books*, September 24, 2015.

Matthews, Chris, "Even Harvard B-School Alums Are Fretting over Income Inequality," *Fortune*, September 9, 2015. As of November 7, 2016: http://fortune.com/2015/09/09/income-inequality-harvard/?iid=sr-link1

May, Ernest R., ed., *American Cold War Strategy: Interpreting NSC 68*, Boston: St. Martin's Press, 1993.

Mazarr, Michael J., "Struggle in the Gray Zone and World Order," *War on the Rocks*, blog, December 22, 2015. As of November 15, 2016: http://warontherocks.com/2015/12/struggle-in-the-gray-zone-and-world-order/

Mazarr, Michael J., and the NDU Strategy Study Group, *Discriminate Power: A Strategy for a Sustainable National Security Posture*, Philadelphia: Foreign Policy Research Institute, May 2013. As of November 8, 2016: http://www.fpri.org/docs/Discriminate_Power_Mazarr_et_al.pdf

McCarthy, Justin, "Americans Remain Upbeat About Foreign Trade," Gallup, February 26, 2016. As of November 8, 2016: http://www.gallup.com/poll/189620/americans-remain-upbeat-foreign-trade.aspx?g_source=trade&g_medium=search&g_campaign=tiles

McCausland, Jeffrey D., *The Post 'Post-Cold War Era' in Europe*, Carlisle, Pa.: Strategic Studies Institute, U.S. Army War College, April 14, 2014.

McFaul, Michael, "How Brexit Is a Win for Putin," *Washington Post*, June 27, 2016.

McKay, Betsy, "The Death Rate is Rising for Middle-Aged Whites," *Wall Street Journal*, November 3, 2015.

McLeod, Gary, George Nacouzi, Paul Dreyer, Mel Eisman, Myron Hura, Krista Langeland, David Manheim, and Geoffrey Torrington, *Enhancing Space Resilience Through Non-Materiel Means*, Santa Monica, Calif.: RAND Corporation, RR-1067-AF, 2016. As of November 2, 2016: http://www.rand.org/pubs/research_reports/RR1067.html

Mead, Walter Russell, *Special Providence: American Foreign Policy and How it Changed the World*, New York: Taylor and Francis Group, 2009.

———, "Grand Strategy: The End of History Ends," *American Interest*, December 2, 2013.

Mearsheimer, John J., "America Unhinged," *The National Interest*, January 2, 2014.

Meredith, Lisa S., Cathy D. Sherbourne, Sarah J. Gaillot, Lydia Hansell, Hans V. Ritschard, Andrew M. Parker, and Glenda Wrenn, *Promoting Psychological Resilience in the U.S. Military*, Santa Monica, Calif.: RAND Corporation, MG-996-OSD, 2011. As of November 2, 2016:
http://www.rand.org/pubs/monographs/MG996.html

Mewett, Christopher, "Understanding War's Enduring Nature Alongside Its Changing Character," *War on the Rocks*, blog, January 21, 2014. As of November 7, 2016:
http://warontherocks.com/2014/01/understanding-wars-enduring-nature-alongside-its-changing-character/

Miller, Paul D., "Organizing the National Security Council: I Like Ike's," *Presidential Studies Quarterly*, Vol. 43, No. 3, September 2013.

Molloy, Raven, Christopher L. Smith, Riccardo Trezzi, and Abigail Wozniak, "Understanding Declining Fluidity in the U.S. Labor Market," Brookings Paper on Economic Activity Conference Draft, March 10–11, 2016. As of November 17, 2016:
https://www.brookings.edu/wp-content/uploads/2016/03/molloytextspring16bpea.pdf

Morrell, Michael, and Bill Harlow, *The Great War of Our Time: The CIA's Fight Against Terrorism from al Qa'ida to ISIS*, New York: Twelve, 2015.

Morris, Errol, and Robert S. McNamara, interview, *The Fog of War*, Sony Pictures Home Entertainment, 2005.

"Mr. Y." (CAPT Wayne Porter, U.S. Navy, and Col Mark "Puck" Mykleby, U.S. Marine Corps), *A National Strategic Narrative*, Washington, D.C.: Woodrow Wilson Center, 2011.

Murray, Charles, *Coming Apart: The State of White America 1960–2010*, New York: Crown Forum, 2012.

Nakashima, Ellen, "Russian Hackers Targeted Arizona Election System," *Washington Post*, August 29, 2016.

National Center for Education Statistics, "Digest of Education Statistics," Table 318.10, "Degrees Conferred by Postsecondary Institutions, by Level of Degree and Sex of Student: Selected Years, 1869–70 through 2024–25," web page, April 2015. As of November 8, 2016:
http://nces.ed.gov/programs/digest/d14/tables/dt14_318.10.asp

National Centers for Environmental Information, "March 11, 2011, Japan Earthquake and Tsunami," web page, March 2015. As of November 8, 2016:
www.ngdc.noaa.gov/hazard/data/publications/2011_0311.pdf

National Endowment for the Arts, "The Arts and Economic Growth," press release, February 16, 2016.

National Science Board, "Science and Engineering Indicators 2014," National Science Foundation, Arlington, Va., NSB 14-01, February 2014.

National Science Foundation, *Federal Funds for Research and Development Detailed Historical Tables: Fiscal Years 1956–1994*, Bethesda, Md., NSF 94-331, 1994.

NATO—*See* North Atlantic Treaty Organization.

Naumann, Klaus, John Shalikashvili, Lord Inge, Jacques Lanxade, Henk van den Breemen, Benjamin Bilski, and Douglas Murray, *Towards a Grand Strategy for an Uncertain World: Renewing Transatlantic Partnership*, Lunteren, Netherlands, Noaber Foundation, 2007.

Navarro, Andrea, "Billionaire Immigrants Who Struck it Rich in the U.S.," Forbes, blog, March 19, 2014. As of November 8, 2016:
http://www.forbes.com/sites/andreanavarro/2014/03/19/billionaire-immigrants-who-struck-it-rich-in-the-u-s/#69759c40c119

Nechepurenko, Ivan, and Neil MacFarquhar, "As Sabotage Blacks Out Crimea, Tatars Prevent Repairs," *New York Times*, November 23, 2015.

New York City Special Initiative for Rebuilding and Resiliency, *PlaNYC: A Stronger, More Resilient New York*, June 2013.

Nordhaus, William D., "A New Solution: The Climate Club," *New York Review of Books*, June 4, 2015.

North Atlantic Treaty Organization, "NATO Leaders Bolster Collective Deterrence and Defence," press release, July 8, 2016.

Nye, Joseph S., Jr., *Bound to Lead: The Changing Nature of American Power*, New York: Basic Books, 1990.

———, *Is the American Century Over?* Cambridge, Mass.: Polity Press, 2015, pp. 123–124.

O'Hanlon, Michael, *U.S. Defense Strategy and the Defense Budget*, Washington, D.C.: Brookings Institution, November 2015.

Obama, Barack, remarks at the National Defense University, Washington, D.C., May 23, 2013.

———, remarks to the United Nations General Assembly, New York, September 28, 2015.

Ochmanek, David, Andrew R. Hoehn, James T. Quinlivan, Seth G. Jones, and Edward L. Warner, *America's Security Deficit: Addressing the Imbalance Between Strategy and Resources in a Turbulent World*, Santa Monica, Calif.: RAND Corporation, RR-1223-RC, 2015. As of October 28, 2016:
http://www.rand.org/pubs/research_reports/RR1223.html

OECD—*See* Organisation for Economic Co-operation and Development.

Office of the Director of National Intelligence, *Global Trends 2030: Alternative Worlds*, Washington, D.C.: National Intelligence Council, 2012.

Office of the Press Secretary, "Fact Sheet: The Nuclear Security Summits: Securing the World from Nuclear Terrorism," The White House, Washington, D.C., March 29, 2016.

Office of Public Affairs, "Launch of Strong Cities Network to Strengthen Community Resilience against Violent Extremism," Washington, D.C.: Department of Justice, September 28, 2015.

Office for the Under Secretary of Defense (Comptroller), *National Defense Budget for FY 17* (Green Book), Washington, D.C.: U.S. Department of Defense, March 2016. As of November 8, 2016:
http://comptroller.defense.gov/Budget-Materials/

Organisation for Economic Co-operation and Development, "Better Life Index," web page, undated-a. As of November 8, 2016:
http://www.oecdbetterlifeindex.org

———, "Income Inequality and Poverty," web page, undated-b. As of November 8, 2016:
http://www.oecd.org/social/inequality-and-poverty.htm

———, "Short-Term Labour Market Statistics: Harmonized Unemployment Rate," database, undated-c. As of November 8, 2016:
http://stats.oecd.org/index.aspx?queryid=36324

———, *In It Together: Why Less Inequality Benefits All*, Paris: OECD Publishing, May 21, 2015. As of November 8, 2016:
http://www.oecd-ilibrary.org/employment/in-it-together-why-less-inequality-benefits-all_9789264235120-en

OriGene Technologies, "CRISPR Cas9 Starter Kit," web page, undated. As of June 27, 2016:
http://www.origene.com/CRISPR-CAS9/Starter-kit.aspx

Overhold, William H., "China and Globalization," testimony to the U.S. China Economic and Security Review Commission, May 19, 2005.

"Overview of Robotic Exoskeleton Suits for Limb Movement Assist," Smashing Robotics, May 6, 2016. As of November 8, 2016:
http://www.smashingrobotics.com/complete-robotic-exoskeleton-suits-list-for-limb-movements/

Panagariya, Arvind, *India's Economic Reforms: What Has Been Accomplished? What Remains to Be Done?* Manila, Philippines: Asian Development Bank, ERDC Policy Brief No. 2, November 2001. As of November 8, 2016:
http://www.columbia.edu/~ap2231/Policy%20Papers/OPB2.pdf

Parks, Bradley C., and Zachary J. Rice, "Does the 'MCC Effect' Exist? Results from the 2012 MCA Stakeholder Survey," *MCA Monitor*, Center for Global Development, February 2013.

Pasquali, Valentina, "The World's Richest and Poorest Countries," *Global Finance*, web page, November 1, 2015. As of May 24, 2016:
https://www.gfmag.com/global-data/economic-data/worlds-richest-and-poorest-countries

Paterson, Lt. Comm. Pat, "The Truth About Tonkin," *Naval History Magazine*, U.S. Naval Institute, Annapolis, Md., Vol. 22, No. 1, February 2008.

Pavel, Barry, and Peter Engelke, *Dynamic Stability: US Strategy for a World in Transition*, Washington, D.C.: Atlantic Council, April 2015.

Perez-Arce, Francisco, Ernesto F. L. Amaral, Haijing Huang, and Carter C. Price, *Inequality and Opportunity: The Relationship Between Income Inequality and Intergenerational Transmission of Income*, Santa Monica, Calif.: RAND Corporation, RR-1509-RC, 2016. As of May 23, 2016:
http://www.rand.org/pubs/research_reports/RR1509.html

Permanent Mission of China to the WTO, "China in the WTO: Past, Present and Future," briefing, 2011.

Petraeus, David, and Paras D. Bhayani, *The Next Great Emerging Market? Capitalizing on North America's Four Interlocking Revolutions*, Cambridge, Mass.: Belfer Center, Harvard Kennedy School, June 2015.

Pew Research Center, *Beyond Distrust: How Americans View Their Government*, Washington, D.C., November 23, 2015a. As of November 8, 2016:
http://www.people-press.org/2015/11/23/beyond-distrust-how-americans-view-their-government/

———, *The American Middle Class Is Losing Ground: No Longer the Majority and Falling Behind Financially*, Washington, D.C., December 9, 2015b. As of November 8, 2016:
http://www.pewsocialtrends.org/files/2015/12/2015-12-09_middle-class_FINAL-report.pdf

———, *Most Americans Say Government Doesn't Do Enough to Help Middle Class*, Washington, D.C., February 4, 2016a. As of November 8, 2016:
http://www.pewsocialtrends.org/2016/02/04/most-americans-say-government-doesnt-do-enough-to-help-middle-class/

———, *Public Uncertain, Divided Over America's Place in the World*, Washington, D.C., May 5, 2016b. As of October 19, 2016:
http://www.people-press.org/2016/05/05/public-uncertain-divided-over-americas-place-in-the-world/

Pezard, Stephanie, and Timothy Smith, "Friends If We Must: Russia and China in the Arctic," *War on the Rocks*, May 6, 2016. As of November 18, 2016:
http://warontherocks.com/2016/05/friends-if-we-must-russia-and-chinas-relations-in-the-arctic/

Pezard, Stephanie, and Abbie Tingstad, "Keep It Chill in the Arctic: Continued U.S. Cooperation in the Arctic Is Key, Despite Simmering Tensions with Moscow," *US News*, April 27, 2016. As of November 18, 2016:
http://www.usnews.com/opinion/articles/2016-04-27/will-the-arctic-remain-a-warm-spot-in-chilly-russia-us-relations

Piccone, Ted, *Five Rising Democracies and the Fate of the Liberal Democratic Order*, Washington, D.C.: Brookings Institution Press, 2016.

Pierce, Justin R., and Peter K. Schott, *The Surprisingly Swift Decline of U.S. Manufacturing Employment*, Cambridge, Mass.: National Bureau of Economic Research, Working Paper No. 18655, December 2012, revised January 2014.

Pistono, Federico, "Robots Will Steal Your Job, But That's OK: How to Survive the Economic Collapse and Be Happy," *Business & Economics*, September 10, 2014.

Popper, Nathaniel, and Guilbert Gates, "Rebuilding Our Infrastructure," *New York Times*, November 14, 2016. As of November 22, 2016:
http://www.nytimes.com/interactive/2016/11/15/business/dealbook/dealbook-infrastructure.html?_r=0

Posen, Barry R., *Restraint: A New Foundation for U.S. Grand Strategy*, Ithaca, N.Y.: Cornell University Press, May 2014.

Posen, Barry R., and Andrew L. Ross, "Competing Visions for U.S. Grand Strategy," *International Security*, Vol. 21, No. 3, Winter 1996/1997, pp. 5–53.

"Putin's Core Support Begins to Waver," Bloomberg, June 9, 2016.

Rabasa, Angel, "Where Are We in the 'War of Ideas'?" in Brian Michael Jenkins and John Paul Godges, eds., *The Long Shadow of 9/11: America's Response to Terrorism*, Santa Monica, Calif., RAND Corporation, MG-1107-RC, 2011. As of December 30, 2016:
http://www.rand.org/pubs/monographs/MG1107.html

Rabasa, Angel, Steven Boraz, Peter Chalk, Kim Cragin, Theodore W. Karasik, Jennifer D. P. Moroney, Kevin A. O'Brien, and John E. Peters, *Ungoverned Territories: Understanding and Reducing Terrorism Risks*, Santa Monica, Calif.: RAND Corporation, MG-561-AF, 2007. As of October 28, 2016:
http://www.rand.org/pubs/monographs/MG561.html

RAND Corporation, "Community Resilience," web page, undated-a. As of November 2, 2016:
http://www.rand.org/topics/community-resilience.html

———, "Infrastructure Resilience and Environmental Policy," web page, undated-b. As of November 2, 2016: http://www.rand.org/jie/infrastructure-resilience-environment.html

———, "Resilience in Action," web page, undated-c. As of November 2, 2016: http://www.rand.org/multi/resilience-in-action.html

"Responsibility to Protect: Two Case Studies with Alex Vines," U.N. Regional Information Centre for Western Europe, undated.

Rice, Susan, and Stewart Patrick, *Index of State Weakness in the Developing World*, Washington, D.C.: Brookings Institution, 2008.

Ries, Charles P., *Improving Decisionmaking in a Turbulent World*, Santa Monica, Calif.: RAND Corporation, PE-192-RC, 2016. As of October 25, 2016: http://www.rand.org/pubs/perspectives/PE192.html

Rodin, Judith, *The Resilience Dividend: Being Strong in a World Where Things Go Wrong*, New York: Public Affairs, 2014.

Rodriguez-Oreggia, Eduardo, *The Impact of Natural Disasters on Human Development and Poverty at the Municipal Level in Mexico*, Cambridge, Mass.: Center for International Development, Harvard University, Working Paper 43, January 2010.

Rodriguez-Oreggia, Eduardo, Alejandro De La Fuente, Rodolfo De La Torre, and Hector A. Moreno, "Natural Disasters, Human Development and Poverty at the Municipal Level in Mexico," *Journal of Development Studies*, Vol. 49, No. 3, March 2013, pp. 442–455.

Romaniuk, Peter, *Does CVE Work? Lessons Learned from the Global Effort to Counter Violent Extremism*, Global Center on Cooperative Security, September 2015.

Roser, Max, "GDP Growth over the Last Centuries," OurWorldInData.org, website, undated-a. As of March 25, 2016: http://ourworldindata.org/data/growth-and-distribution-of-prosperity/gdp-growth-over-the-last-centuries/

———, "Political Regime," OurWorldInData.org, website, undated-b. As of November 18, 2016: https://ourworldindata.org/grapher/political-regime?logo=DRI&year=1930

Roser, Max, and Esteban Ortiz-Ospina, "Literacy," OurWorldInData.org, website, 2016. As of February 18, 2015: http://ourworldindata.org/data/education-knowledge/literacy/

Rothkopf, David, *National Insecurity: American Leadership in an Age of Fear*, New York: Public Affairs, 2014.

Rumelt, Richard P., *Good Strategy, Bad Strategy*, New York: Crown Publishing Group, 2011.

"Russia Unveils New Navy Icebreaker in Arctic Military Focus," *DefenseNews*, June 11, 2016. As of November 8, 2016:
http://www.defensenews.com/story/defense/naval/2016/06/11/russia-unveils-new-navy-icebreaker-arctic-military-focus/85747556/

Samuelson, Robert J., "It's Still the Economy, Stupid," *Washington Post*, February 3, 2016.

Schelling, Thomas C., *Arms and Influence*, Princeton, N.J.: Princeton University Press, 1966.

Schiavenza, Matt, "Why the U.S. Is Stuck with Saudi Arabia: Even the Shale Oil Revolution Can't Wean Washington off its Despotic Middle Eastern Ally," *The Atlantic*, January 2015.

Schwabish, Jonathan, and Courtney Griffith, "The U.S. Federal Budget: A Closer Look at Discretionary Spending," fact sheet, Congressional Budget Office, Washington, D.C., April 2012. As of June 7, 2016:
https://www.cbo.gov/sites/default/files/BS_Discretionary_print.pdf

Scott, Mark, "Companies Struggle to Make Carbon Capture Viable," *New York Times*, October 5, 2015. As of November 17, 2016:
http://www.nytimes.com/2015/10/06/business/energy-environment/companies-struggle-to-make-carbon-capture-viable.html?_r=0

Sestanovich, Stephen, *Maximalist: America in the World from Truman to Obama*, New York: Knopf, 2014.

Sharif, Abdullah, "China's Ambitions in Asia," *World Post*, November 13, 2015. As of November 8, 2016:
http://www.huffingtonpost.com/abdullah-sharif/chinas-ambitions-in-asia_b_8538504.html

Shatz, Howard J., *U.S. International Economic Strategy in a Turbulent World*, Santa Monica, Calif.: RAND Corporation, RR-1521-RC, 2016. As of October 27, 2016:
http://www.rand.org/pubs/research_reports/RR1521.html

Shaw, Dave, "The Economy's Improving but Americans' Economic Anxiety Persists," *Marketplace*, March 14, 2016. As of November 8, 2016:
http://www.marketplace.org/2016/03/11/economy/anxiety-index/economys-improving-americans-economic-anxiety-persists

Shimizu, Mika, and Allen L. Clark, "Interconnected Risks, Cascading Disasters and Disaster Management Policy: A Gap Analysis," planet@risk, Vol. 3, No. 2, 2015. As of November 8, 2016:
https://planet-risk.org/index.php/pr/article/view/199/408

Shlapak, David A., "Toward a More Modest American Strategy," *Survival: Global Politics and Strategy*, Vol. 57, No. 2, March 20, 2015, pp. 59–78.

Shlapak, David A., and Michael Johnson, *Reinforcing Deterrence on NATO's Eastern Flank: Wargaming the Defense of the Baltics*, Santa Monica, Calif.: RAND Corporation, RR-1253-A, 2016. As of November 06, 2016: http://www.rand.org/pubs/research_reports/RR1253.html

Shultz, George P., *Issues on My Mind: Strategies for the Future*, Stanford, Calif.: Hoover Institution Press, 2013.

Singer, Peter W., Heather Messera, and Brendan Orino, *D.C.'s New Guard: What Does the Next Generation of American Leaders Think?* Brookings Institution, February 2011.

Slaughter, Anne-Marie, "Does Obama Have a Grand Strategy for his Second Term? If Not, He Could Try One of These," *Washington Post*, January 18, 2013.

Smeltz, Dina, *Foreign Policy in the New Millennium: Results of the 2012 Chicago Council Survey of American Public Opinion and U.S. Foreign Policy*, Chicago, Ill.: Chicago Council on Global Affairs, 2012.

Smeltz, Dina, Ivo H. Daalder, Karl Friedhoff, and Craig Kafura, "America in the Age of Uncertainty," Chicago, Ill.: Chicago Council on Global Affairs, October 2016.

Smith, Julianne, "Our Overworked Security Bureaucracy," *Democracy Journal*, No. 40, Spring 2016.

Smith, Katherine F., Michael Goldberg, Samantha Rosenthal, Lynn Carlson, Jane Chen, Cici Chen, and Sohini Ramachandran, "Global Rise in Human Infectious Disease Outbreaks," *Journal of the Royal Society Interface*, Vol. 11, No. 101, October 29, 2014. As of November 8, 2016: http://rsif.royalsocietypublishing.org/content/11/101/20140950.full

Smithsonian National Museum of American History, "Western Indian Wars," web page, undated. As of June 22, 2016: http://amhistory.si.edu/militaryhistory/printable/section.asp?id=6

Snyder, Glenn, *Deterrence and Defense*, Princeton, N.J.: Princeton University Press, 1961.

"Spring 2015 Global Attitudes Survey," Pew Research Center, web page, Washington, D.C., June 23, 2015. As of November 8, 2016: http://www.pewglobal.org/2015/06/23/spring-2015-survey/

Stares, Paul B., *The Preventive Imperative: How America Can Avoid War, Stay Strong, and Keep the Peace in the 21st Century*, New York: Colombia University Press, forthcoming, 2017.

State Council Information Office of the People's Republic of China, *China's Military Strategy*, Beijing, May 2015.

Stewart, Lt. Gen. Vincent, director of the Defense Intelligence Agency, testimony to the U.S. House of Representatives Foreign Affairs Committee, March 2, 2016.

Stimson Center, *Strategic Agility: Strong National Defense for Today's Global and Fiscal Realities*, Washington, D.C., 2013.

Stockholm International Peace Research Institute, "Global Nuclear Weapons: Downsizing but Modernizing," web page, June 13, 2016. As of June 13, 2016: https://www.sipri.org/media/press-release/2016/global-nuclear-weapons-downsizing-modernizing

Stork, Joe, "The Carter Doctrine and U.S. Bases in the Middle East," *Middle East Report*, Vol. 10, September/October 1980.

Strachan, Hew, "Strategy and Contingency," *International Affairs*, Vol. 87, No. 16, 2011, pp. 1281–1296.

Stuster, J. Dana, "Russian Security Now Using Typewriters to Thwart the NSA," *Foreign Policy*, July 11, 2013. As of June 24, 2016: http://foreignpolicy.com/2013/07/11/russian-security-now-using-typewriters-to-thwart-the-nsa/

Sullivan, Erin M., Joseph L. Annest, Feijun Luo, Thomas R. Simon, and Linda L. Dahlberg, "Suicide Among Adults Aged 35-64 Years—United States, 1999–2010," Centers for Disease Control and Prevention, *Morbidity and Mortality Weekly Report*, Vol. 62, No. 17, May 3, 2013, pp. 321–325.

Tabuchi, Hiroko, "Inquiry Declares Fukushima Crisis a Man-Made Disaster," *New York Times*, July 5, 2012.

Taleb, Nassim Nicholas, "Learning to Love Volatility," *Wall Street Journal*, November 16, 2012.

Tavernise, Sabrina, "Life Spans Shrink for Least Educated Whites in the U.S.," *New York Times*, September 20, 2012.

Thompson, Clive, *Smarter Than You Think: How Technology Is Changing Our Minds for the Better*, New York: Penguin, 2013.

Thrall, A. Trevor, and Erik Goepner, *Millennials and U.S. Foreign Policy*, Cato Institute, June 16, 2015.

Trump, Donald J., "Donald Trump's Contract with the American Voter," web page, November 10, 2016. As of November 18, 2016: https://assets.donaldjtrump.com/_landings/contract/O-TRU-102316-Contractv02.pdf

UN—*See* United Nations.

United Nations, *Millennium Development Goals Report*, New York: Department of Economic and Social Affairs, 2015a. As of February 22, 2016: http://www.un.org/millenniumgoals/2015_MDG_Report/pdf/MDG%202015%20rev%20(July%201).pdf

———, *World Population Prospects, 2015 Revision*, New York: Department of Economic and Social Affairs, 2015b.

United Nations Regional Centre for Peace and Disarmament in Asia and the Pacific, "Biological Weapons Convention Meeting of Experts to Convene in Geneva from 10 to 14 August 2015," web page, August 7, 2015. As of October 3, 2016: http://unrcpd.org/news-item/biological-weapons-convention-meeting-of-experts-to-convene-in-geneva-from-10-to-14-august-2015/

United Nations Regional Information Centre for Western Europe, "Responsibility to Protect: Two Case Studies with Alex Vines," web page, undated. As of September 8, 2016: http://www.unric.org/en/responsibility-to-protect/26982-responsibility-to-protect-two-case-studies-with-alex-vines

United Nations Security Council Resolution 1894, on protection of civilians in armed conflict, November 11, 2009.

United Nations Security Council Resolution 1975, on Cote D'Ivoire, March 30, 2011.

USAID—*See* U.S. Agency for International Development.

U.S. Agency for International Development, Center of Excellence on Democracy, Human Rights, and Governance, *Stand with Civil Society: Best Practices*, January 2014. As of November 8, 2016: https://www.usaid.gov/sites/default/files/documents/1866/SWCS%20Best%20Practices%20Publications%20Final%20(1).pdf

U.S. Bureau of Labor Statistics, "Current Employment Statistics," database, series ID CES5051110001 and series ID CES5051700001, undated. As of April 11, 2016: http://data.bls.gov/timeseries/CES0000000001 http://data.bls.gov/timeseries/CES5051700001?data_tool=XGtable

———, "The Employment Situation, September 2016," news release, September 2, 2016.

U.S. Census Bureau, "Intercensal Estimates of the United States Resident Population by Age and Sex," Washington, D.C., 1991.

———, "Historical Poverty Tables: People and Families—1959 to 2015," web page, Table 20, "Poverty Status of Related Children Under 6 Years of Age," September 2016a. As of October 1, 2016: https://www.census.gov/data/tables/time-series/demo/income-poverty/historical-poverty-people.html

———, "Income, Poverty and Health Insurance Coverage in the United States: 2015," News Release CB16-158, September 13, 2016b.

"U.S. Concern About Global Warming at Eight-Year High," Gallup poll, web page, March 16, 2016. As of November 10, 2016: http://www.gallup.com/poll/190010/concern-global-warming-eight-year-high.aspx

U.S. Department of State, *Congressional Budget Justification, Department of State, Foreign Operations and Related Programs, Fiscal Year 2015*, Washington, D.C., March 4, 2014.

———, *Trafficking in Persons Report*, Washington, D.C., July 2015. As of November 8, 2016:
http://www.state.gov/documents/organization/245365.pdf

U.S. Department of State, Bureau of Democracy, Human Rights and Labor, *Russia 2013 Human Rights Report*, Washington, D.C., 2013. As of November 8, 2016:
http://www.state.gov/documents/organization/220536.pdf

"U.S. Position in the World," Gallup, web page, February 3–7, 2016. As of February 5, 2016:
http://www.gallup.com/poll/116350/position-world.aspx

Vaidyanathan, Gayathri, "Thinning Arctic Ice Proves Deadly for Veteran Explorers," *Scientific American*, May 6, 2015. As of November 2, 2016:
https://www.scientificamerican.com/article/
thinning-arctic-ice-proves-deadly-for-veteran-explorers1/

van Agtmael, Antoine, and Fred Bakker, *The Smartest Places on Earth: Why Rustbelts Are the Emerging Hotspots of Global Innovation*, New York: Public Affairs, 2016.

Victor, D. G., J. House, and S. Joy, "A Madisonian Approach to Climate Policy," *Science*, Vol. 309, 2005, pp. 1820–1821.

Walker, Christopher, "The Authoritarian Threat: The Hijacking of Soft Power," *Journal of Democracy*, Vol. 27, No. 1, January 2016. As of November 18, 2015:
http://www.ned.org/wp-content/uploads/2016/01/
January-2016-JOD-Hijacking-of-Soft-Power-Christopher-Walker.pdf

Walt, Stephen M., "Is America Addicted to War?" *Foreign Policy*, April 4, 2011.

Watts, Jonathan, and Jan Rocha, "Brazil President Weeps as She Unveils Report on Military Dictatorship's Abuses," *The Guardian*, December 10, 2014. As of November 8, 2016:
https://www.theguardian.com/world/2014/dec/10/
brazil-president-weeps-report-military-dictatorship-abuses

Weiner, Eric, "The Secret of Immigrant Genius," *Wall Street Journal*, January 15, 2016.

"When is Enough . . . Enough? Why the Wealthy Cannot Get Off the Treadmill," UBS Investor Watch, 2Q 2015. As of June 3, 2016:
http://view.ceros.com/ubs/investor-watch-treadmill/p/1

White House, *National Security Strategy of the United States of America*, Washington, D.C., September 2002.

———, *United States Mid-Century Strategy for Deep Decarbonization*, November 2016a.

———, "Statement by the President on Actions in Response to Russian Malicious Cyber Activity and Harassment," Washington, D.C., December 29, 2016b.

Wike, Richard, and Katie Simmons, "Global Support for Principle of Free Expression, but Opposition to Some Forms of Speech," Pew Research Center, web page, November 18, 2015. As of November 8, 2016:
http://www.pewglobal.org/2015/11/18/global-support-for-principle-of-free-expression-but-opposition-to-some-forms-of-speech/

"William S. Burroughs, Quotable Quote," goodreads.com, undated. As of November 1, 2016:
https://www.goodreads.com/quotes/134938-the-best-way-to-keep-something-bad-from-happening-is

Williams-Grut, Oscar, "Robots Will Steal Your Job: How AI Could Increase Unemployment and Inequality," *Business Insider*, February 15, 2016. As of November 8, 2016:
http://www.businessinsider.com/robots-will-steal-your-job-citi-ai-increase-unemployment-inequality-2016-2

Willis, Henry H., and Kathleen Loa, *Measuring the Resilience of Energy Distribution Systems*, Santa Monica, Calif.: RAND Corporation, RR-883-DOE, 2015. As of November 2, 2016:
http://www.rand.org/pubs/research_reports/RR883.html

Wood, Nathan J., Jeanne Jones, Seth Spielman, and Mathew C. Schmidtlein, "Community Clusters of Tsunami Vulnerability in the US Pacific Northwest," *Proceedings of the National Academy of Sciences*, Vol. 112, No. 17, March 15, 2015, pp. 5354–5359.

Work, Bob, Deputy Secretary of Defense, speech at National Defense University, Washington, D.C., August 5, 2014.

———, speech at the Reagan Presidential Library, Simi Valley, Calif., November 7, 2015.

Worland, Justin, "Paris Climate Change Agreement Set to Take Effect After Quick Ratification Process," Time.com, October 5, 2016. As of October 12, 2016:
http://time.com/4519895/paris-agreement-ratification-european-union/

World Bank, "Managing Disaster Risks for Resilient Development," April 10, 2014. As of November 8, 2016:
http://www.worldbank.org/en/results/2013/04/12/managing-disaster-risks-resilient-development

———, *Russia Economic Report 34: Balancing Economic Adjustment and Transformation*, September 30, 2015a. As of November 8, 2016:
http://www.worldbank.org/en/news/press-release/2015/09/30/russia-economic-report-34

———, "Overview," web page, October 7, 2015b. As of April 13, 2016: http://www.worldbank.org/en/topic/poverty/overview

———, *Russian Federation Monthly Economic Developments*, November 11, 2015c.

———, "World Development Indicators," database, February 17, 2016a. As of May 29, 2016: http://data.worldbank.org/data-catalog/world-development-indicators

———, "GDP Ranking," web page, April 11, 2016b. As of June 2016: http://data.worldbank.org/data-catalog/GDP-ranking-table

———, *High and Dry: Climate Change, Water and the Economy*, Washington, D.C., May 2016c.

World Economic Forum, *The Global Risks Report 2016*, Geneva, 11th edition, 2016.

"World University Rankings, 2015–16," Times Higher Education, London, undated.

Worstall, Tim, "Trump Doesn't Need to Bother Apple About Manufacturing in America—Foxconn Replaces 60,000 With Robots," Forbes, May 27, 2016. As of November 18, 2016: http://www.forbes.com/sites/timworstall/2016/05/27/trump-neednt-bother-apple-about-manufacturing-in-america-foxconn-replaces-60000-with-robots/#34a27f2f58a2

Zarate, Juan, "Treasury's War: The Unleashing of a New Era of Financial Warfare," *Public Affairs*, New York, 2013.

Zenko, Micah, *Red Team: How to Succeed by Thinking Like the Enemy*, New York: Basic Books, 2015.